Chemistry Olympiad
Support Booklet

written by

Phil Copley

Tim Hersey

Chas McCaw

Rob Paton

Kathryn Scott

Andrew Worrall

Peter Wothers

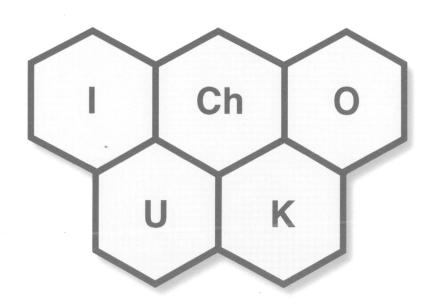

Chemistry Olympiad Support Booklet

Written by:
Phil Copley, Bishop Wordsworth's School, Salisbury
Tim Hersey, Harrow School, Middlesex
Chas McCaw, Winchester College, Winchester
Rob Paton, University of Cambridge
Kathryn Scott, Oxford University
Andrew Worrall, Harrow School, Middlesex
Peter Wothers, University of Cambridge

Edited by Emma Woodley
Designed by Russel Spinks

Published and distributed by the Royal Society of Chemistry
Printed by the Royal Society of Chemistry

ISBN-13: 978-1-84755-866-4

British Library Cataloguing in Publication Data.

A catalogue for this book is available from the British Library.

Acknowledgements

The Chemistry Olympiad Committee would like to thank the following people and organisations for their help, advice and support in the preparation of this booklet.

The experts from JAWS anthill for the factual information on ants and for the photograph of the Formica Rufa worker ant. *www.anthillwood.co.uk* (accessed 08/07/08).

The photography department of the University of Cambridge, Department of Chemistry for the photographs of sherbet lemons, the halon fire extinguisher, Fexofenadine, and copper sulfate.

The NMR dept of the University of Cambridge, Department of Chemistry for running the ^{19}F NMR spectrum for the question about the spectra of haloalkanes, and the 700 MHz ^{1}H NMR spectrum of NanoBalletDancer.

Professor James M. Tour and his research group for the kind loan of a sample of NanoBalletDancer.

Anthony Lim, a student at the University of Cambridge, for the cartoon used in the rat poison question.

The cone snail image is courtesy of Dr. Baldomero Olivera, photograph by K. S. Matz.

The UK Chemistry Olympiad is supported by:

INEOS is a leading global manufacturer of petrochemicals, specialty chemicals and oil products. It comprises 19 businesses each with a major chemical company heritage. The production network spans 73 manufacturing facilities in 19 countries throughout the world. The chemicals INEOS produce enhance almost every aspect of modern life.

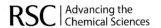
RSC | Advancing the Chemical Sciences

Foreword

INEOS has great pleasure in supporting the UK selection process for the annual International Chemistry Olympiad. This challenging competition provides students with the opportunity to use their knowledge in new and sometimes unfamiliar contexts, whilst developing their problem solving skills. This booklet contains a selection of typical questions from Round 1 of the competition, each followed by a detailed discussion of how the answers can be derived. We hope that this booklet will help ambitious students and their teachers prepare for the Olympiad, and provide a source of engaging material for all.

Jim Ratcliffe
Chairman, INEOS

Contents

RSC | Advancing the Chemical Sciences

Introduction

Every year the Royal Society of Chemistry (RSC) organises the selection of the UK team for the International Chemistry Olympiad (IChO). The IChO has been running for 40 years, and the UK has been involved since 1983. Next year, in July 2009, the UK will be hosting the competition, and almost 300 students from about 70 countries will be attending.

This booklet is designed with two purposes in mind: to help and encourage ambitious post-16 chemistry students to do well in future competitions; and as a valuable teaching resource for schools to stretch and challenge gifted students.

The UK selection process is open to UK Sixth Form students, with Lower Sixth as well as Upper Sixth students encouraged to take part. Students entering the competition compete in two rounds. The first round is a written paper taken at the students' own institution, and the second is a two day event held at a university. The students most successful in the second round will compose the team that will travel to the international final.

The Round 1 paper is a written test of chemical knowledge and understanding. A small committee of teachers from schools and universities spends a great deal of time and effort coming up with what we think are interesting and challenging questions on real and relevant chemistry, raising awareness of what the subject is all about. Tackling the Round 1 paper also provides a good opportunity to develop some of the skills required for study at university and beyond. There is no doubt that these questions are demanding: they do not rely on the relatively easy recall of information which students will have met before, but instead on thinking and trying to work out answers to unfamiliar questions – this is much more difficult.

The recent Round 1 questions, which are contained in this booklet, are typical; they may seem 'impossible' at first sight, but the authors have tried to show how it is possible to work out the answers, sometimes by seeing analogies to what they have done, sometimes piecing data together rather like a jigsaw puzzle. The style of each solution is unashamedly varied and reflects the different approaches of individual teachers. Success is often achieved by sheer determination and not giving up. We hope that this booklet will enable students to prepare in the most effective way, and help them to approach the Round 1 paper with confidence.

Later rounds of the Olympiad competition involve support and teaching from a number of truly inspirational teachers who guide the eventual team, hopefully towards medals but certainly towards a great deal of fun. But the first hurdle for potential Chemistry Olympians is to excel in Round 1. Be ambitious and enter!

More past papers and answers can be found on the RSC website *www.rsc.org/olympiad*.

Good luck!

Tim Hersey	Emma Woodley
Chairman	Assistant Education Manager
UK Chemistry Olympiad Selection Committee	Royal Society of Chemistry

Sherbet Lemons

From Round 1 2006, Question 1.

Sherbet lemons are sweets which consist of a flavoured sugar shell filled with sherbet.

The sherbet contains sodium hydrogencarbonate and tartaric acid (2,3-dihydroxybutanedioic acid).

(a) Assuming all the sugar present is sucrose, $C_{12}H_{22}O_{11}$, write an equation for the complete combustion of the sugar.

(b) The standard enthalpy change of combustion of sucrose is -5644 kJ mol^{-1}. Calculate the energy released when one sweet containing 6.70 g of sucrose is completely burnt.

(c) A man needs to consume about 2500 dietary calories per day. Given that 1 kJ $\equiv$ 0.239 dietary calories, how many sweets must a man consume in order to meet his daily calorific requirement?

Sherbet produces a slight fizzing sensation in the mouth when the tartaric acid reacts with the sodium hydrogencarbonate to make carbon dioxide. In a laboratory experiment, one sherbet lemon sweet produced 6.00 cm^3 of carbon dioxide.

(d) Calculate the minimum masses of tartaric acid and sodium hydrogencarbonate necessary to produce this volume of carbon dioxide.
    [Assume 1 mole of any gas occupies 24.0 dm^3 at r.t.p.]

A carbon atom bonded to four different groups is called a chiral centre or an asymmetric carbon atom. A molecule which contains just one chiral centre exists as two stereoisomers (isomers containing the same groups attached to the same atoms). These stereoisomers are non-superimposable mirror images of each other called *enantiomers*. If a molecule contains more than one chiral centre, the number of stereoisomers increases and some of the stereoisomers may be superimposable on their mirror images.

(e) By making the appropriate substitutions for **a**, **b**, **c**, and **d** in the structure shown below, draw all the different stereoisomers of tartaric acid, indicating clearly which (if any) are enantiomers.

$$HO_2C \diagdown \qquad \diagup CO_2H$$
$$b^{\text{''''''}}C - C^{\text{'''''}}c$$
$$a \qquad\qquad d$$

RSC | Advancing the Chemical Sciences

(f) Citric acid is used to flavour sherbet lemons. Its formula may be written $HOOCCH_2.C(OH)(COOH).CH_2COOH$. How many asymmetric carbon atoms does this molecule contain?

Sherbet Lemons – an analysis

The first question on each Olympiad paper should be really approachable, to get students 'in the mood' for the more demanding questions to come. This one is very general, with the content of most parts not much beyond GCSE level; the more advanced material is introduced and explained, so good Lower Sixth students should be able to do well. We would like to encourage more Lower Sixth students to enter this competition, both to interest and enthuse them, and as an opportunity to practise in readiness for another attempt next year.

(a) The equation should be fairly easy to write once the products CO_2 and H_2O are identified.

$$C_{12}H_{22}O_{11} + 12\ O_2 \rightarrow 12\ CO_2 + 11\ H_2O$$

(b) The moles of sucrose ($M_r = 342$) present in a sherbet lemon is $6.70 / 342$, so the energy released is $(6.70 / 342) \times 5644 = 111$ kJ.

(c) This calculation should encourage sensible students not to eat too many sweets! 111 kJ is equivalent to $111 \times 0.239 = 26.5$ dietary calories, so 2500 dietary calories could come from $2500 / 26.5 =$ just under 95 sweets.

(d) The question starts to get slightly trickier now. Students have to realise from the systematic name of tartaric acid that it produces two acidic H^+ ions. Students might attempt to write a fully balanced chemical equation for the reaction taking place, but it is simpler just to realise that 1 mole of tartaric acid will react with 2 moles of $NaHCO_3$ to produce 2 moles of CO_2. When set out logically the calculation for part (d) is as follows:

$6.00 / 24000 = 2.50 \times 10^{-4}$ moles of CO_2 produced,

so the mass of tartaric acid ($M_r = 150$) $= 1/2 \times 2.50 \times 10^{-4} \times 150 = 0.0188$ g,

and the mass of sodium hydrogencarbonate ($M_r = 84$) $= 2.50 \times 10^{-4} \times 84 = 0.0210$ g.

(e) Lower Sixth students and some Upper Sixth students are unlikely to have covered stereoisomerism so some important ideas are explained in the question. This is a typical technique which we use to test thinking rather than just factual recall in Olympiad questions; many well prepared A-level students find this difficult, however.

After reading this paragraph they should realise the importance of drawing molecules using the appropriate three dimensional representation, and the necessity of the absence of a plane of symmetry in a molecule if it is to display optical isomerism. Drawing the structure of tartaric acid clearly might suggest that there are four stereoisomers (since there are two asymmetric carbon atoms present which can each have 'left' and 'right' handed orientation), but since the chiral centres are identical there are in fact only three. These are shown below. The two mirror-image molecules which are non-superimposable on each other are enantiomers (optical isonomers). The third isomer contains the plane of symmetry and so is not optically active; it is said to be achiral.

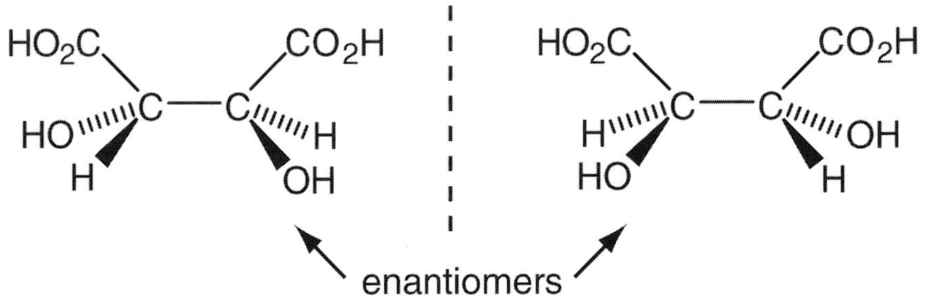

enantiomers

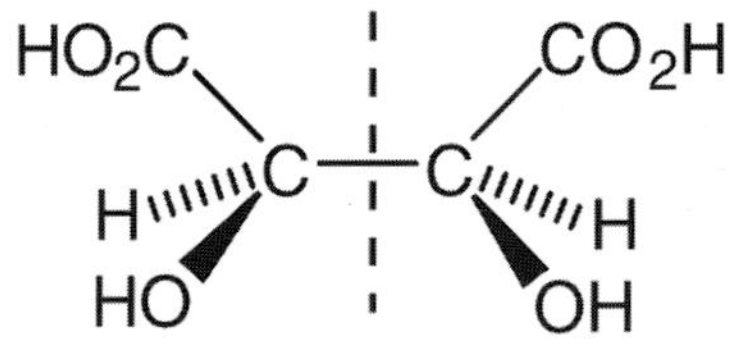

not optically active
(achiral)

(f) If the structure of citric acid is drawn clearly from the condensed structure given it is apparent that it contains no chiral centres: the central C contains two identical CH_2COOH groups. Candidates who have not given up already should be rewarded with the mark.

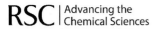

Ants

From Round 1 2005, Question 3.

The 'simplest' carboxylic acid is called methanoic acid and has formula HCOOH. It occurs naturally in ants and used to be prepared by distilling them! This gave rise to the earlier name for methanoic acid – *formic* acid – after the Latin word *formica* for ant.

When an ant bites, it injects a solution containing 50% by volume of methanoic acid. A typical ant may inject around 6.0×10^{-3} cm^3 of this solution.

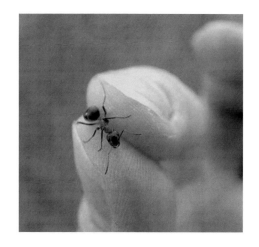

A *Formica rufa* worker ant, just after biting the photographer

(a) i) When you are bitten by an ant it does not inject you with all of its methanoic acid but keeps a little in reserve. Assuming a 'typical ant' injects 80% of its methanoic acid, what is the total volume of pure methanoic acid contained in a 'typical ant'?

ii) How many 'typical ant' ants would have to be distilled to produce 1.0 dm^3 of pure methanoic acid?

Bicarbonate of soda (sodium hydrogencarbonate) is often used to treat ant stings.

(b) i) Write the equation for the reaction between sodium hydrogencarbonate and methanoic acid.

ii) Given that the density of methanoic acid is 1.2 g cm^{-3}, how many moles of methanoic acid does the 'typical ant' inject?

iii) What mass of sodium hydrogencarbonate would be needed to neutralise completely the sting from this ant?

(c) As soon as the methanoic acid is injected it dissolves in water in the body to produce a solution of methanoic acid. Assuming that it dissolves immediately in 1.0 cm^3 of water in the body, calculate the concentration of the methanoic acid solution that is formed. [You may ignore the volume of the methanoic acid itself in this calculation.]

The pH of a solution is related to the concentration of hydrogen ions as follows:
$$pH = -\log[H^+]$$
where [H$^+$] stands for the concentration of hydrogen ions in mol dm^{-3}.

(d) The pH of the methanoic acid solution produced above was 2.43. What is the concentration of hydrogen ions in this solution?

Methanoic acid is a weak acid and so is only partly ionised in solution
$$HCOOH(aq) \rightleftharpoons HCOO^-(aq) + H^+(aq)$$

(e) Calculate the percentage of methanoic acid molecules which are ionised in this solution.

The *acid dissociation constant*, K_a, is a measure of how ionised a weak acid is. For methanoic acid it is defined by the following expression, where again square brackets written round a formula mean the concentration of that substance in mol dm^{-3}
$$K_a = [HCOO^-][H^+] / [HCOOH].$$

(f) Calculate the acid dissociation constant for methanoic acid.

Ants – an analysis

This is another example of a 'starter' question, this one on acids. Some question parts involve straightforward calculations, while the later parts have the more advanced topics explained so that they should be accessible to all students who are prepared to think.

(a) i) A 'typical ant' should contain $6.0 \times 10^{-3} \times 0.5 \times 100 / 80 = 3.75 \times 10^{-3}$ cm^3 of pure methanoic acid; since the data is only given to two significant figures the answer should be given as 3.8×10^{-3} cm^3.

   ii) The number of 'typical ants' = $1000 / (3.75 \times 10^{-3}) = 2.7 \times 10^5$. It is worthwhile for students to consider whether or not their answer is realistic: many answers submitted were obviously incorrect, and ranged from 2.67 (watch how you pick up one of these!) via 266667 to 6.4 million without very much common sense or feel for significant figures.

(b) i) The equation for the reaction between sodium hydrogencarbonate and methanoic acid is likely to be unfamiliar, but it is basically the same as other acid / carbonate reactions.

   $$NaHCO_3 + HCOOH \rightarrow HCOONa + H_2O + CO_2$$

   ii) The ant injects $(6.0 \times 10^{-3} \times 0.5)$ cm^3 pure methanoic acid; this has a mass of $(6.0 \times 10^{-3} \times 0.5 \times 1.2)$ g so represents $6.0 \times 10^{-3} \times 0.5 \times 1.2 / 46 = 7.8 \times 10^{-5}$ moles of methanoic acid.

   iii) The moles calculation now follows: we need 7.8×10^{-5} moles of $NaHCO_3 = 7.8 \times 10^{-5} \times 84 = 6.6 \times 10^{-3}$ g or 6.6 mg. Again this seems reasonable.

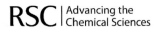

(c) The concentration of methanoic acid is obviously 7.8×10^{-5} moles in 1.0 cm^3 or 0.0010 dm^3 of solution, *ie* $7.8 \times 10^{-5} / 0.0010 = 7.8 \times 10^{-2}$ mol dm^{-3}.

(d) Students may not have calculated pH before so the formula was given; they had to use their calculator properly to determine the H$^+$ concentration as 3.7×10^{-3} mol dm^{-3}.

(e) The percentage of HCOOH molecules ionising is therefore:
$(3.7 \times 10^{-3} / 7.8 \times 10^{-2}) \times 100 = 4.8\%$. This shows how little weak acids ionise.

(f) The final part of the calculation was the trickiest; students had to calculate the acid dissociation constant for methanoic acid from the formula given. The concentration of HCOO$^-$ ions is obviously equal to that of the H$^+$ ions. The concentration of undissociated HCOOH is no longer 7.8×10^{-2} mol dm^{-3} since some of these molecules are ionising; it is reduced by those which ionised, *ie* 3.7×10^{-3} mol dm^{-3}. So the acid dissociation constant $K_a = (3.7 \times 10^{-3})^2 / ((7.8 \times 10^{-2}) - (3.7 \times 10^{-3})) = 1.8 \times 10^{-4}$.

Answers in the range $1.8 - 1.9 \times 10^{-4}$, with or without units, were given full credit.

Mercury fulminate

From Round 1 2008, Question 3.

Mercury(II) fulminate, $HgC_2N_2O_2$, has been known as a super-sensitive explosive for 300 years, but – being so difficult to handle – its crystal structure was only determined in 2007. To avoid setting off an explosion, the compound had to be synthesised in the dark in a process that the researchers describe as "quite tricky".

On detonation mercury(II) fulminate decomposes forming three products: two of these are gases, two are elements.

(a) Write the equation for the detonation of mercury(II) fulminate.

(b) Calculate, using an energy cycle or otherwise, the standard enthalpy change of reaction for this detonation using the following standard enthalpies of formation, $\Delta_fH^{\ominus}$.

[Note that not all of the data are required.]

Compound	$HgC_2N_2O_2$	HgO	CO	CO_2	NO	NO_2
$\Delta_fH^{\ominus}$ / kJ mol^{-1}	+386	−91	−111	−394	+90	−33

Mercury(II) fulminate can be described as *organometallic*, meaning that it has a metal-carbon bond. The fulminate ion is a triatomic ion with a charge of minus one. The infrared spectra of fulminates include a bond stretch in the triple-bond region.

(c) Suggest a structure for mercury(II) fulminate, showing the number and type of bonds between the atoms.

Mercury(II) cyanate is an isomer of mercury(II) fulminate. In fact cyanates and fulminates were the first known examples of isomers in chemistry. Mercury(II) cyanate is not organometallic but its infrared spectrum does contain a bond stretch in the triple-bond region.

(d) Suggest the structure of mercury(II) cyanate, showing the number and type of bonds between the atoms.

Protonation of cyanates yields cyanic acid, which isomerises to form isocyanic acid, HNCO. This then spontaneously trimerises to form cyanuric acid, $(HNCO)_3$. There are two different structures for cyanuric acid which both exist in equilibrium. In both structures the three

atoms for any given element occupy symmetrically equivalent positions; one structure appears to be aromatic.

(e) Showing all bonds, draw the two possible structures for cyanuric acid.

Mercury fulminate – an analysis

Students are not expected to have any knowledge of the fulminate or cyanate ions or any of the related acids. They need to apply logically the information in the question, as well as doing an energy cycle calculation, and using the valencies of carbon, nitrogen and oxygen to suggest structures.

(a) It is important to realise here that the decomposition of mercury fulminate on detonation does not involve reaction with oxygen or anything else – just pure decomposition. We are told that two of the products are elements. These are likely to be the most unreactive elements in the compound if they are to remain uncombined after detonation. Nitrogen is well known to be inert and is a likely candidate. (Indeed most explosives contain nitrogen so that N_2 is produced, with its very exothermic bond formation.) Mercury is an unreactive metal; carbon and oxygen are more chemically reactive, and are commonly chemically bound together. The preamble also says that two of the products are gases. Assuming nitrogen is one of them, that is consistent with the carbon and oxygen being combined as carbon monoxide (a gas) and mercury (a liquid) as the other product. These conclusions point to the following equation:

$$HgC_2N_2O_2 \rightarrow Hg + N_2 + 2CO$$

(b) This is a typical enthalpy cycle question except the data in the question also gives some quantities that are not needed. (If only the data for $HgC_2N_2O_2$ and CO were provided, that would have given away the answer to part (a).) The elements are not quoted as their standard enthalpies of formation are zero by definition. In general the enthalpy change for a reaction is the sum of the enthalpy changes of formation for the products minus the enthalpy changes of formation for the reactants. Since the elements have a standard enthalpy change of formation of zero, the enthalpy change of reaction is:

$$\Delta_r H^{\ominus}$$
$$HgC_2N_2O_2(s) \longrightarrow Hg(l) + N_2(g) + 2CO(g)$$

$\Delta_f H^{\ominus}(HgC_2N_2O_2)$ $2 \times \Delta_f H^{\ominus}(CO)$

$$Hg(l) + N_2(g) + 2\,C(s) + O_2(g)$$

$$\Delta_r H^{\ominus} = -\Delta_f H^{\ominus}(HgC_2N_2O_2) + (2 \times \Delta_f H^{\ominus}(CO))$$

$$= -386\ kJ\ mol^{-1} + (2 \times (-111\ kJ\ mol^{-1})) = -608\ kJ\ mol^{-1}$$

Note that this is a very exothermic reaction considering that it is just a decomposition reaction (which are commonly endothermic). This fact and the gas released in the reaction are consistent with mercury fulminate being a good explosive.

(c) If the fulminate ion is triatomic, then mercury(II) fulminate must be $Hg(CNO)_2$ and the fulminate ion must be $(CNO)^-$. We are told there is a metal–carbon bond; it is likely therefore to be a single bond as the ion is singly charged, with the negative ion charge formally on the carbon. So mercury(II) fulminate contains the C–Hg–C group; the mercury–carbon bonds have significant covalent character. The infrared spectrum indicates that there is a triple bond. Since carbon has a valency of 4 this suggests that the fulminate ion is linear with the carbon triply bonded to the nitrogen (as nitrogen has a valency of 3). Since this triple bond satisfies nitrogen's valency its bond to oxygen must be a dative bond. As the dative bond involves two electrons from the nitrogen then that satisfies the valency of the oxygen. The structure can be written with an arrow representing the dative bond or as a normal covalent bond combined with formal charges from an electron transfer:

$$O \leftarrow N \equiv C - Hg - C \equiv N \rightarrow O \quad \text{or} \quad {}^-O - N^+ \equiv C - Hg - C \equiv N^+ - O^-$$

(d) As the cyanate ion contains a triple bond as well, it very probably also contains a $C \equiv N$ triple bond, as there is no obvious way of drawing the ion with a $C \equiv O$ or $N \equiv O$ triple bond. Another way of drawing the ion with a $C \equiv N$ triple bond is to bond the oxygen to the carbon instead. With a single bond from the carbon to the oxygen, satisfying carbon's valency, the oxygen would need to carry the negative charge to satisfy its own valency. It follows that the bond with the mercury ion is through the oxygen atom, which is consistent with mercury(II) cyanate not being organometallic. It therefore has the structure:

$$N \equiv C - O - Hg - O - C \equiv N$$

(e) The structure of isocyanic acid is given as H–N=C=O. The most likely way for this molecule to polymerise will be for linkages between monomers to be through carbon and nitrogen, as having higher valencies than O or H they have more bonds available. They are also multiply bonded in isocyanic acid, which reminds us of the multiply bonded carbons in alkene polymerisation:

If this fragment trimerised into a ring it would have threefold symmetry, putting the atoms of each element into equivalent positions:

RSC | Advancing the Chemical Sciences

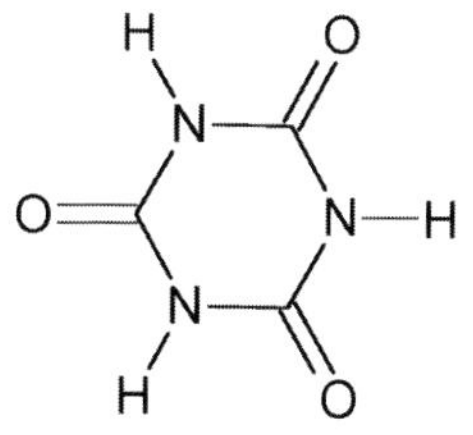

This molecule can be drawn in a different way while preserving the alternate carbon and nitrogen hexagon. The hydrogen could be bonded to the oxygen with the double bond moving into the ring:

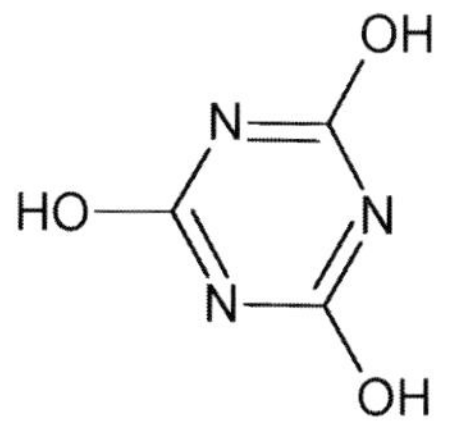

The threefold symmetry is preserved and all the valencies of the atoms are still correct. In this form there are alternate single and double bonds in the ring which is what we associate with the six carbon atoms in the benzene ring. This one is therefore the isomer that is considered aromatic.

Geometric shapes in chemistry

From Round 1 2007, Question 2.

Phosphorus exists as a number of allotropes, the most reactive being white phosphorus. This was first prepared in the 17th century from the reduction of the phosphate present in urine.

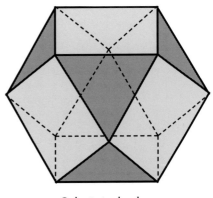

Cubotetrahedron

Solid white phosphorus contains P_4 molecules, with each P atom at the vertex of a regular tetrahedron.

(a) Draw the structure of a molecule of P_4, showing all the chemical bonds.

(b) How many edges are there in total in a regular tetrahedron?

White phosphorus spontaneously ignites in air to form a mixture of phosphorus(III) oxide and phosphorus(V) oxide.

(c) Write an equation for the formation of i) phosphorus(III) oxide and ii) phosphorus(V) oxide from the elements.

The structure of each oxide is also based on a regular tetrahedron. The phosphorus atoms remain at the vertices but are no longer bonded to each other. Instead the P atoms are joined by bridging oxygens.

(d) Draw the structure of phosphorus(III) oxide.

Phosphorus(V) oxide has a further oxygen atom bonded to each phosphorus atom at the vertex of the tetrahedron.

(e) Draw the structure of phosphorus(V) oxide.

Each oxide reacts with water to form an acid – phosphorus(V) oxide forms phosphoric(V) acid, H_3PO_4.

(f) Draw the molecular structure of phosphoric(V) acid, showing all of the bonds.

(g) Write the equation for the reaction forming phosphoric(V) acid.

A quantitative method for determining phosphate levels in aqueous solution involves adding ammonium molybdate, $(NH_4)_2MoO_4$, to form a precipitate of ammonium molybdophosphate. The structure of this solid is based on a cuboctahedron (shown above).

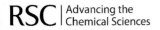
RSC | Advancing the Chemical Sciences

A molybdenum atom lies at each vertex of the cuboctahedron and these are joined by oxygen atoms with every edge of the cuboctahedron being bridged by an oxygen atom. A further oxygen atom is joined to every vertex. A single phosphate unit lies at the centre of the structure with each of its four oxygen atoms coordinating to three molybdenum atoms.

(h) Calculate the oxidation state of molybdenum in ammonium molybdate.

(i) How many i) vertices and ii) edges are there in a cuboctahedron?

(j) Calculate the number of i) molybdenum atoms and ii) oxygen atoms in the molybdophosphate ion.

(k) Given that no atom changes its oxidation state during the formation of ammonium molybdophosphate, calculate the overall charge of the molybdophosphate ion and hence the formula of ammonium molybdophosphate.

Geometric shapes in chemistry – an analysis

This question is to encourage students to think in three dimensions – specifically in the context of regular polyhedra. No knowledge of the chemistry of the substances is required, although there may be some advantage early on if the chemistry of phosphorus has been covered. The only required knowledge is the calculation of oxidation numbers and the shape of the tetrahedron. Students need to interpret the picture of the cuboctahedron and the written descriptions of structures in the question.

(a) This is the shape of a tetrahedron where P is at the corners (vertices) and the edges are P–P single bonds.

Note that the bond angles of 60° are very different from the optimum bond angles; the resulting bond strain accounts for the dangerously reactive nature of white phosphorus, P_4.

(b) The six edges of a tetrahedron are highlighted here as they are relevant to the structure of the oxides of phosphorus coming up in the question. This particular shape is of interest as it is the most symmetrical four-cornered three-dimensional object – the four P atoms at the corner positions are symmetrically equivalent. With all the faces and edges equivalent as well, a tetrahedron is a Platonic solid. Note that the number of faces plus the number of vertices minus the number of edges is equal to two: this is common to all three-dimensional shapes.

(c) Considering the oxidation number of phosphorus in each of the oxides, the formula of the (+III) oxide is P_2O_3 and (+V) oxide is P_2O_5. The next two parts of the question reveal that P_2O_3 and P_2O_5 are in fact only empirical formulae, and that the actual molecular formulae are P_4O_6 and P_4O_{10} respectively. The use of empirical or molecular formulae are both acceptable for the reaction equations, *ie*

$$P_4 + 3\,O_2 \rightarrow 2\,P_2O_3 \quad \text{OR} \quad P_4 + 3\,O_2 \rightarrow P_4O_6$$
$$P_4 + 5\,O_2 \rightarrow 2\,P_2O_5 \quad \text{OR} \quad P_4 + 5\,O_2 \rightarrow P_4O_{10}$$

(d) and (e) The information in the question describes the structures as:

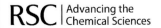

It is the connectivity of the atoms that is the most important for our purposes rather than ensuring that the overall shape looks realistic. As these structures are still based on four phosphorus atoms in a tetrahedral arrangement (though no longer bonded directly to each other), the four P atoms are still equivalent. The six "edge" oxygen atoms are also equivalent. In P_4O_{10} the four P=O atoms are equivalent. The oxidation number indicates the number of bonds to oxygen on each phosphorus; this confirms that the four extra oxygens in P_4O_{10} are doubly bonded to the phosphorus atoms. A dative bond from P to its vertex oxygen also satisfies the valencies of the atoms.

(f) Guidance is provided by the structure of the corresponding oxide molecule, which has one P=O double bond and three P–O single bonds. The three hydrogens in H_3PO_4 could add to the singly-bonded oxygens to make up their valency.

The stereochemistry is not required to be shown in this structure.

(g) As the question has revealed the formulae of phosphorus(V) oxide and phosphoric(V) acid the equation is just a problem of balancing:

$$P_4O_{10} + 6\ H_2O \rightarrow 4\ H_3PO_4$$

(h) Since the ammonium ion is NH_4^+, the molybdate ion must be MoO_4^{2-} so that the charges balance. The oxidation number of the atoms comprising an ion must sum to the charge, and oxygen is assigned an oxidation number of -2 in virtually all of its compounds, so the oxidation number of molybdenum, O.N.(Mo) is found as follows:

$$O.N.(Mo) + (4 \times -2) = -2 \qquad \therefore O.N.(Mo) = +6$$

(i) As the edges of the back side of the cuboctahedron are shown with dotted lines it is easy to count the edges and vertices, giving 24 edges and 12 vertices. These answers can also be arrived at by considering the types of faces involved in the cuboctahedron. The figure reveals that there are six square faces that share vertices but none of their edges. Further inspection reveals that every edge in the cuboctahedron is associated with one of the squares. Each square has four edges so that makes a total of 24 edges in the cuboctahedron. A similar line of reasoning applied to the eight triangular faces shows that there are 24 edges altogether. Interestingly, every vertex connects two square faces and two triangular ones; all 12 vertices are equivalent. Every edge connects a square face with a triangular one; all 24 edges are equivalent. With all the vertices and edges equivalent but with different types of face, the cuboctahedron is an Archimedean solid.

(j) As there are 12 vertices on the cuboctahedron, there are 12 molybdenum atoms in the molybdophosphate ion. An oxygen atom is in the centre of each edge and also one is bonded to each vertex (rather like in P_4O_{10}) making a total of $12 + 24 = 36$ oxygens as part of the cuboctahedron. There are also four oxygens in the phosphate unit at the centre of the cuboctahedron, making 40 oxygens altogether in molybdophosphate.

(k) We have established that the ion has the formula $Mo_{12}PO_{40}^{n-}$. It is likely to be negatively charged to balance the positive ammonium ions and whatever the positive ions were that balanced the phosphate used to make it. We can establish the charge by adding the oxidation numbers of all the atoms
(Mo = +6, P = +5, O = -2).

$$(12 \times 6) + 5 + (40 \times -2) = -3$$

Therefore the formula of ammonium molybdophosphate is $(NH_4)_3Mo_{12}PO_{40}$.

The spectra of haloalkanes

From Round 1 2006, Question 7.

Haloalkanes have been used as aerosol propellants and refrigerants but are now largely banned due to the damage they cause to the ozone layer. Halon 1211 was once commonly used in fire extinguishers (now only found in fighter jets) and 'Halothane' is an inhalational general anaesthetic.

Further examples of haloalkanes are given in the table below.

	Common name	Structural formula
A	CFC-113	$Cl_2FC\text{-}CClF_2$
B	CFC-113a	$Cl_3C\text{-}CF_3$
C	HFC-134a	$F_3C\text{-}CH_2F$
D	CFC-11 (Freon-11, R-11)	CCl_3F
E	CFC-12 (Freon-12, R-12)	CCl_2F_2
F	CFC-13	$CClF_3$
G	Halon 1211	$CBrClF_2$
H	Methylene bromide	CH_2Br_2

Whilst naturally occurring carbon and fluorine exist as essentially the single isotopes ^{12}C and ^{19}F, chlorine consists of 75% ^{35}Cl and 25% ^{37}Cl; bromine consists of 50% ^{79}Br and 50% ^{81}Br. The presence of chlorine and bromine atoms in molecules therefore leads to characteristic patterns for molecular ions in mass spectrometry. As an example, the mass spectrum of CFC-13 (**F**) includes peaks at $m/z = 104$ ($CF_3{}^{35}Cl^{\bullet+}$) and 106 ($CF_3{}^{37}Cl^{\bullet+}$) with intensity ratio 3:1.

(a) Calculate the m/z values and relative intensities for the molecular ion peaks of CFC-12 (**E**).

(b) Sketch the mass spectrum for the molecular ion peaks of Halon 1211 (**G**). Indicate the relative intensity of each peak and which ion(s) are responsible for them.

A sample of methylene bromide (**H**) was enriched with deuterium (^2H). On analysis it was

found that half of the hydrogen content of the sample was deuterium. In the mass spectrum there are molecular ion peaks with m/z values of 172, 173, 174, 175, 176, 177 and 178.

(c) Calculate the relative intensities of these molecular ion peaks.

NMR spectroscopy is a technique which reveals the number of different environments of certain nuclei in a molecule. NMR active nuclei such as ^1H, ^{13}C and ^{19}F are routinely studied. As an example, the two hydrogen atoms in methylene bromide (**H**) are equivalent and hence would give rise to a single peak in the ^1H NMR spectrum. The same is true for the hydrogens in HFC-134a (**C**).

(d) Complete the table below indicating the number of different fluorine environments for the each of the compounds **A–G**.

	Common name	Structural formula	Number of different fluorine environments
A	CFC-113	$Cl_2FC-CClF_2$	
B	CFC-113a	Cl_3C-CF_3	
C	HFC-134a	F_3C-CH_2F	
D	CFC-11 (Freon-11, R-11)	CCl_3F	
E	CFC-12 (Freon-12, R-12)	CCl_2F_2	
F	CFC-13	$CClF_3$	
G	Halon 1211	$CBrClF_2$	

(e) The anaesthetic *Halothane* has the formula $C_2HBrClF_3$ and shows one signal in its ^{19}F NMR spectrum. Draw the *two* possible three-dimensional structures for Halothane.

The intensity of a signal in a ^1H or ^{19}F NMR spectrum is proportional to the number of nuclei in that particular environment.

(f) For each compound with more than one signal in its ^{19}F NMR spectrum, indicate in the appropriate column of the table the expected intensity ratio.

NMR spectra are complicated by coupling between nuclei. If an NMR-active nucleus is within three bonds of another similar nucleus *which is in a different chemical environment*, its signal will be split into a number of peaks instead of appearing as a single peak. If a nucleus couples to n NMR-active nuclei, its signal will split into a total of $(n +1)$ peaks.

(g) The ^{19}F NMR spectrum of one of the haloalkanes from the table is shown overleaf. Draw the structure of the haloalkane and indicate with an arrow which fluorines give rise the signals X and Y.

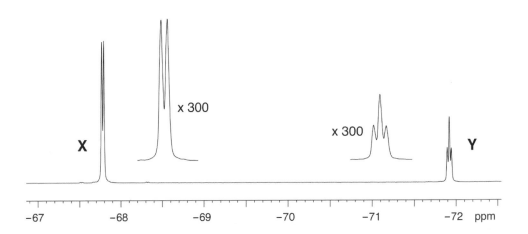

The spectra of haloalkanes – an analysis

This is a challenging mass spectrometry and NMR question because students need to consider mass ratios in the mass spec with more than one element with multiple isotopes, and the NMR involves the unfamiliar fluorine-19 nucleus. However, ^{19}F behaves much like ^{1}H in NMR and so familiar principles can be applied.

(a) With two chlorine atoms in the molecule, each with two possible isotopes, the molecular ion can be made up with four different combinations, though two of them will have the same mass:

$CF_2{}^{35}Cl^{35}Cl^+$ Mass = 12 + (2×19) + (2×35) = 120
$CF_2{}^{35}Cl^{37}Cl^+$ Mass = 12 + (2×19) + 35 + 37 = 122
$CF_2{}^{37}Cl^{35}Cl^+$ Mass = 12 + (2×19) + 37 + 35 = 122
$CF_2{}^{37}Cl^{37}Cl^+$ Mass = 12 + (2×19) + (2×37) = 124

The probability of detecting each species is the product of the probabilities of finding each of the isotopes, where $P(^{35}Cl) = \frac{3}{4}$ and $P(^{37}Cl) = \frac{1}{4}$. Therefore:

$P(CF_2{}^{35}Cl^{35}Cl^+) = \frac{3}{4} \times \frac{3}{4} = 9/16$
$P(CF_2{}^{35}Cl^{37}Cl^+) = P(CF_2{}^{37}Cl^{35}Cl^+) = \frac{3}{4} \times \frac{1}{4} = 3/16$
$P(CF_2{}^{37}Cl^{37}Cl^+) = \frac{1}{4} \times \frac{1}{4} = 1/16$

The ratio of the intensities of the peaks at 120, 122 and 124 is the ratio of the probabilities calculated for each mass, which are conveniently measured above in fractions over 16. Therefore the ratios are 9:6:1 as there are two species, each of probability 3/16, contributing to the 122 peak.

(b) This is much like part (a) with two atoms in the molecule each comprised of two isotopes. The difference is that the two combinations of isotopes with the same mass do not have equal probability:

$CF_2{}^{35}Cl{}^{79}Br^+$ Mass = 164   Probability = ¾ × ½ = 3/8
$CF_2{}^{35}Cl{}^{81}Br^+$ Mass = 166   Probability = ¾ × ½ = 3/8
$CF_2{}^{35}Cl{}^{79}Br^+$ Mass = 166   Probability = ¼ × ½ = 1/8
$CF_2{}^{35}Cl{}^{81}Br^+$ Mass = 168   Probability = ¼ × ½ = 1/8

Adding the probabilities for the two peaks at mass 166, the ratio of the peak intensities at 164, 166 and 168 is therefore 3:4:1. The spectrum therefore should look like:

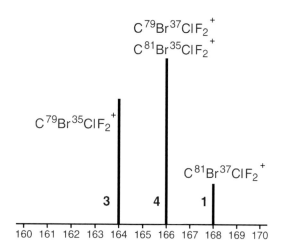

(c) In the sample of methylene bromide, following the deuterium enrichment, all four atoms bonded to the carbon are comprised of two isotopes in a 50:50 ratio – ^1H and ^2H, and ^{79}Br and ^{81}Br. There are therefore 2 × 2 × 2 × 2 = 16 isomers, though some will have the same mass. The lightest will be $C^1H^1H^{79}Br^{79}Br^+$, with a mass of 172, and the heaviest $C^2H^2H^{81}Br^{81}Br^+$, with a mass of 178. Both of these isomers have a probability of ½ × ½ × ½ × ½ = 1/16. In fact, all isomers will have this probability since the probability of any individual isotope being present is ½. That means it is only a question of counting the number of isotopes of each mass.

Considering the isomers with a mass one different from these extremes: this difference in one mass unit must result in the exchange of a hydrogen atom for a deuterium atom, or *vice versa*. As there are two hydrogens in the molecule there will be two ways of exchanging the isotope, doubling the probability of finding an isomer of that mass. The isomer with mass 173 is therefore $C^2H^1H^{79}Br^{79}Br^+$ or $C^1H^2H^{79}Br^{79}Br^+$, while the mass 177 isomer is $C^1H^2H^{81}Br^{81}Br^+$ or $C^2H^1H^{81}Br^{81}Br^+$.

Considering the next unit of mass in from the extremes – 174 and 176 – there are more possibilities. Being two units of mass from the extremes, this can be achieved either by exchanging both hydrogen atoms for deuterium atoms (or *vice versa*) or by changing one of the Br isotopes. There are two ways of changing a bromine isotope (as there are two bromines in the molecule) but only one way of changing *both* hydrogen isotopes. This makes a total of three isomers at each of the masses 174 and 176.

Finally there is the isotope of mass 175, which is three mass units in from the extremes. Increasing the mass by three units from the lightest isomer can only be achieved by using a heavier isotope of both a bromine and a hydrogen (a similar argument can be used about lighter isotopes with the heaviest isomer). However there is a choice of bromine atom and a choice of hydrogen atom, so there are four possible isomers for the mass 175 peak. All the isomers are collected in the table:

Mass	Isomers	Relative Intensity
172	$C^1H^1H^{79}Br^{79}Br^+$	1
173	$C^2H^1H^{79}Br^{79}Br^+$, $C^1H^2H^{79}Br^{79}Br^+$	2
174	$C^2H^2H^{79}Br^{79}Br^+$, $C^1H^1H^{81}Br^{79}Br^+$, $C^1H^1H^{79}Br^{81}Br^+$	3
175	$C^2H^1H^{81}Br^{79}Br^+$, $C^2H^1H^{79}Br^{81}Br^+$, $C^1H^2H^{81}Br^{79}Br^+$, $C^1H^2H^{79}Br^{81}Br^+$	4
176	$C^1H^1H^{81}Br^{81}Br^+$, $C^2H^2H^{79}Br^{81}Br^+$, $C^2H^2H^{81}Br^{79}Br^+$	3
177	$C^1H^2H^{81}Br^{81}Br^+$, $C^2H^1H^{81}Br^{81}Br^+$	2
178	$C^2H^2H^{81}Br^{81}Br^+$	1

The relative intensity is just the number of possible isomers for a given mass, all 16 isomers have a 1/16 probability.

(d) Where there are multiple fluorine atoms bonded to a carbon atom, they will be in the same environment. In the case of a CF_3 group, they will be able to rotate into each other's position (single bonds can generally freely rotate) and rotation is fast on the NMR time scale (the spectrum is a time-average over bond rotations); in a CF_2 group the two F atoms are symmetrically equivalent as there is a mirror plane bisecting them. Therefore there is only one fluorine environment for the haloalkanes **B**, **D**, **E**, **F** and **G**. In **A** and **C** fluorine atoms are bonded to different carbon atoms that are not made equivalent by symmetry so in these cases there are two fluorine environments.

	Common name	Structural formula	Number of different fluorine environments
A	CFC-113	$Cl_2FC\text{-}CClF_2$	2
B	CFC-113a	$Cl_3C\text{-}CF_3$	1
C	HFC-134a	$F_3C\text{-}CH_2F$	2
D	CFC-11 (Freon-11, R-11)	CCl_3F	1
E	CFC-12 (Freon-12, R-12)	CCl_2F_2	1
F	CFC-13	$CClF_3$	1
G	Halon 1211	$CBrClF_2$	1

(e) Given that there is only one signal in the ^{19}F NMR spectrum, the three fluorine atoms in *halothane* must be bonded to the same carbon. So it seems there can only be one isomer, but consideration of the other carbon reveals that it has four different groups bonded to it (H, Br, Cl and CF_3), making it a chiral centre. There are therefore two optical isomers that are mirror images of each other. Dot-and-wedge bonds are consequently required to distinguish between the isomers:

(f) This is just a question of identifying the number of fluorine atoms bonded to each carbon in molecules **A** and **C** where there is more than one fluorine environment. The ratio is therefore 1:2 in **A** and 3:1 in **C**.

(g) There are two fluorine signals in the spectrum (**X** and **Y**). Therefore the molecule can only be **A** or **C**, as only these have two fluorine environments. In **A**, since there is no hydrogen atom, the splitting of the peaks can only be due to other fluorine atoms (the carbon and chlorine atoms are not NMR-active). The Cl_2FC fluorine signal can be expected to be split into three peaks (*ie* a triplet) since there are two other fluorine atoms within three bonds and in a different chemical environment. The ClF_2C fluorine signal can be expected to be split into two peaks (*ie* a doublet) since there is one other fluorine atom within three bonds and in a different chemical environment. The predicted doublet and triplet are the two features in the spectrum in the question, and the intensities of the peaks are also consistent with the 1:2 ratio of the fluorine atoms in the environments of **Y** and **X** respectively in **A**.

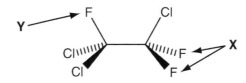

C is not a possible answer to this question. The presence of the hydrogen atoms, which are NMR-active, would complicate the splitting considerably. Even neglecting the effect of the hydrogens, the spectrum of **C** would still feature a signal split into four peaks from coupling with CF_3.

The NMR spectra of NanoPutians

From Round 1 2005, Question 4.

In June 2003, a research paper was published announcing the synthesis of the smallest representations of the human form: 2 nm tall anthropomorphic molecules, nicknamed 'NanoPutians' by their creators.

The molecules synthesised included 'NanoKid', 'NanoBaker' and 'NanoAthlete'. The compound shown to the right was called 'NanoBalletDancer' and has the formula $C_{41}H_{50}O_2$.

When assigning an NMR spectrum, the first step is to identify how many atoms there are in unique environments.

Both carbon atoms (^{13}C) and hydrogen atoms (^1H) give NMR signals. Each atom *in a different environment* will give rise to one signal.

For example, in the structure of NanoBalletDancer, carbon atoms 37 and 39 are equivalent; we may write (37 ≡ 39). Hence, although there are two carbon atoms (37 and 39) which have one oxygen atom attached, only *one* signal would be observed in a ^{13}C NMR spectrum due to these carbon atoms since they are equivalent.

(a) Which carbon atoms making up the benzene rings are equivalent? Write down w ≡ x, y ≡ z etc for any equivalent atoms. How many signals *in total* would be observed due to benzene-ring carbons in a carbon NMR spectrum of NanoBalletDancer?

(b) List the groups of equivalent triple bond carbons in NanoBalletDancer. How many signals would be seen *in total* in the ^{13}C spectrum due to triple bond carbon atoms?

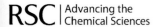

(c) How many different methyl groups (–CH$_3$ groups) are there in NanoBalletDancer? Again, list them in groups.

(d) How many different carbon environments are there in NanoBalletDancer – *ie* how many signals would be seen in total in the ^{13}C NMR spectrum?

Similarly, in ^{1}H NMR, the total number of signals depends on the number of different environments of hydrogen atoms in a structure. There are 13 different environments of hydrogens in NanoBalletDancer; their signals are labelled **A–M** in the spectrum below. The numbers of hydrogen atoms in each unique environment is given under the label. Hydrogen atoms in similar environments all have similar chemical shifts. For example, all the hydrogens on the benzene rings occur in the same region of the spectrum, *ie* they have a similar *chemical shift*.

However, ^{1}H NMR is complicated by *coupling*. If a hydrogen is within three bonds of another hydrogen *which is in a different environment*, instead of appearing as a single peak, its signal is split into a number of peaks. In general, if the hydrogen under consideration is within three bonds of *n* hydrogens in a different environment from the one under consideration, it will be split into (*n* + 1) peaks. The ratio of the area under the peaks is given by Pascal's triangle as outlined below.

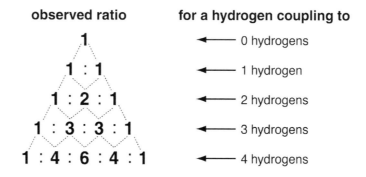

The signal for a given hydrogen is not split by any hydrogens *which are in the same environment* as it is in.

(e) Into how many peaks will the signal from a hydrogen that couples with 5 other hydrogens be split? What will the ratio of the peaks be?

It is possible to assign the ^{1}H NMR spectrum of NanoBalletDancer by considering the numbers of hydrogens in different environments, their chemical shifts, and their coupling patterns. For example, the signal at 7.15 ppm (**B**) is due to the hydrogen atoms on carbons 19 and 23.

(f) On the table overleaf, assign (as far as possible) which signals are due to which hydrogen atoms. The assignment for signal **B** has already been filled in. (For some signals, it might not be possible to decide between two alternative assignments – in which case just write '... *or* ...')

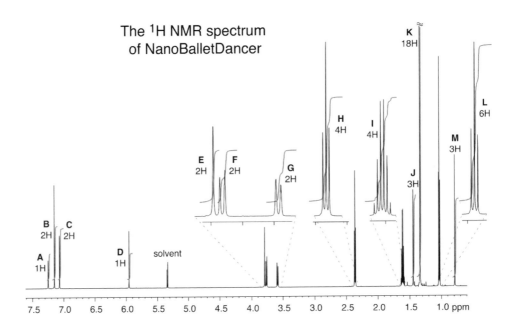

The ^{1}H NMR spectrum of NanoBalletDancer

^{1}H NMR signal	Hydrogen(s) on Carbon(s)
A	
B	19, 23
C	
D	
E	
F	
G	
H	
I	
J	
K	
L	
M	

The NMR Spectra of NanoPutians – an analysis

This question was designed to get students thinking about the symmetry present in molecules. This is a key idea needed when interpreting NMR spectra. Since everyone is familiar with the symmetry present in the human form, the NanoPutian molecules seemed an ideal way to introduce the idea.

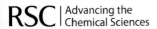

The question starts with ^{13}C NMR since this is so much more easily understood than ^{1}H NMR. There are no complications from coupling to worry about. All that is needed is an appreciation of the symmetry and hence numbers of different environments of ^{13}C nuclei.

It is highly recommended to construct a model in order to answer this question. As the figure is drawn in the question, the dancer is doing a high kick. It may appear that the symmetry has been lost. However, it must be remembered that it is possible to rotate about the single bonds. Rotating about the single bonds in the waist of the molecule gives a structure in which the symmetry is obvious. The two structures shown below show this rotation.

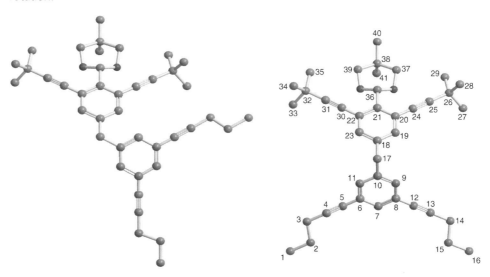

(a) There are two benzene rings in the structure – the upper body, and the lower body. Note that the double bonds are fully delocalized around the ring so we do not need to worry about where they appear in the structure.

Considering the lower body first:
Atom 7 is unique, 6 ≡ 8, 9 ≡ 11, and 10 is unique. The lower body therefore gives rise to four signals in the ^{13}C spectrum.

For the upper body:
Atom 18 is unique, 19 ≡ 23, 20 ≡ 22, and 21 is unique. The upper body therefore also gives rise to four signals.

In total there will be **eight** signals in the ^{13}C spectrum due to the benzene-ring carbons.

(b) The triple bonds are in the arms and legs of the structure.

For the legs: 4 ≡ 13, 5 ≡ 12

For the arms: 24 ≡ 30, 25 ≡ 31

In total there will be **four** signals in the ^{13}C spectrum due to the triple bond carbons.

(c) The methyl groups on the feet are equivalent: 1 ≡ 16

All the finger methyl groups are equivalent: 27 ≡ 28 ≡ 29 ≡ 33 ≡ 34 ≡ 35

The slightly tricky point concerns the two methyl groups on the head, 40 and 41. These are different. One of these, 41, is on the same side of the ring that makes up the head as the body (both go into the plane of the paper). The other, 40 is on the opposite side of the head-ring as the body. Even though the ring that is the head may flip, these methyl groups will never be equivalent. This is shown in the diagram below.

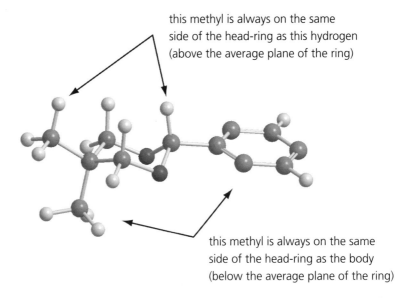

this methyl is always on the same side of the head-ring as this hydrogen (above the average plane of the ring)

this methyl is always on the same side of the head-ring as the body (below the average plane of the ring)

This can be tricky to appreciate but is exactly the same problem as realising the two compounds shown below are different.

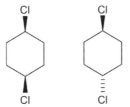

These two compounds cannot be inter-converted without breaking bonds.

This means there are **four** different environments of methyl carbons in the structure: the toes, the fingers, and the two on the head.

(d) In total, there will be 23 signals in the ^{13}C NMR spectrum corresponding to the following groups:

RSC | Advancing the Chemical Sciences

1	1 ≡ 16	15	24 ≡ 30
2	2 ≡ 15	16	25 ≡ 31
3	3 ≡ 14	17	26 ≡ 32
4	4 ≡ 13	18	27 ≡ 28 ≡ 29 ≡ 33 ≡ 34 ≡ 35
5	5 ≡ 12	19	36 unique
6	7 unique	20	37 ≡ 39
7	6 ≡ 8	21	38 unique
8	9 ≡ 11	22	40 unique
9	10 unique	23	41 unique
10	17 unique		
11	18 unique		
12	19 ≡ 23		
13	20 ≡ 22		
14	21 unique		

The second part of the question looks at proton NMR. This is complicated by coupling between different environments of protons. However, when trying to assign a ^{1}H NMR spectrum, the first thing to do is to work out how many different environments of protons there are in order to predict how many separate signals should be seen in the spectrum.

We are told in the question that there are 13 different environments of protons in the structure, but it is worthwhile to list these, together with how many protons are in each environment. Since the number of protons making up each signal are given in the spectrum, we can go a long way in assigning the spectrum without considering any coupling at all.

Protons on carbon number	Number of protons protons in this environment	Possible signal in spectrum (based on number of protons alone)
1 and 16	6 (three on each)	L
2 and 15	4 (two on each)	H or I
3 and 14	4 (two on each)	H or I
7	1	A or D
9 and 11	2 (one on each)	B, C, E, or F
17	2	B, C, E, or F
19 and 23	2 (one on each)	B, C, E, or F
27, 28, 29, 33, 34, 35	18 (three on each)	L
36	1	A or D
37 upper and 39 upper	2 (one on each)	B, C, E, or F
37 lower and 39 lower	2 (one on each)	B, C, E, or F
40	3	J or M
41	3	J or M

To make the final assignments, we do need to consider the coupling patterns.

The information about coupling is given in the question. We are only concerned about interactions between different protons up to three bonds apart.

(e) Continuing Pascal's triangle tells us that a proton that couples with five other equivalent protons will split a signal into six peaks (a sextet) with ratio 1:5:10:10:5:1.

This sextet is actually the pattern observed for signal **I**. The two equivalent hydrogens on carbon 15 are split by the five hydrogens within three bonds – three on carbon 16, and two on carbon 14. Note that the hydrogens on carbon 16 are not the same as those on carbon 14, but the coupling interactions are sufficiently close so a clear sextet of signals is seen in the NMR signal of **I**.

(f) Signal **L** arises from the three hydrogens on carbon 16 and their equivalent counterparts on carbon 1. These hydrogens couple with the two hydrogens on carbon 15, and the two on carbon 2 respectively. Coupling to two equivalent signals splits the signal into the triplet seen in the spectrum with ratio 1:2:1.

From the discussion in part (e), the four hydrogens which give rise to signal **I** are the two on carbon 15, and their equivalent pair on carbon 2.

The other signal in the spectrum which arises from four equivalent hydrogens is **H**. This is a clear triplet, which means the equivalent hydrogens must be coupling with just two other hydrogens. The hydrogens on carbon 14 only couple to the two hydrogens on carbon 15 and so give rise to a triplet with ratio 1:2:1.

The hydrogens on carbon 14 are equivalent to the pair on carbon 3, and together these four give rise to signal **H** in the spectrum.

It is not easy to decide between the singlets **A** and **D** (which are due to one hydrogen each), or between the singlets **B** and **C** (which are due to two hydrogens each). The only hydrogens which could give rise to 1H singlets are the hydrogen on carbon 36 (the carbon with two oxygens attached), and the hydrogen on carbon 7 (an aromatic carbon).

The only hydrogens which give rise to the 2H singlets are the two hydrogens on carbon 17 and those on the aromatic carbons 9 (and its counterpart 11) and 19 (and its counterpart 23).

It might be sensible to expect all the signals due to hydrogens in similar environments to have similar shifts, and this is the case. All the hydrogens on aromatic carbons occur between 7.0 and 7.3 ppm in the spectrum. Signal **B** is already assigned as being due to two of the hydrogens on the body benzene ring and signal **A** is also due to a hydrogen on a benzene ring - carbon 7. Signal **D** is therefore due to the

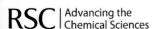

hydrogen on carbon 36. The two hydrogens on carbon 17, which is not an aromatic carbon, give rise to the signal **E** in the spectrum.

To distinguish between signals **B** and **C** is harder still, and this was not necessary for full marks. In fact in the question, **B** is already assigned. However, for completeness, we shall tell you how it is possible to distinguish between these signals. There is a difference in the appearance of these two singlets: **B** is almost twice as tall as **C** even though they are both due to two hydrogens. The reason **C** is not as tall is because there is a very small coupling interaction between the hydrogens on carbon 9 (and 11) with the hydrogen on carbon 7. These hydrogens are separated by four bonds, so the coupling interaction is much weaker than usual and not really noticeable in the spectrum. However, whilst the coupling does not split the signal **C** completely into two, the interaction means the signal is broader and this explains why it is shorter. The areas under signals **B** and **C** are the same but **B** is tall and thin, whereas the weak coupling interaction means that **C** is shorter and broader.

This same weak coupling interaction also broadens signal **A** and indeed, this signal is shorter and broader than signal **D** which is tall and thin.

There are two pairs of signals which we are not able to firmly assign from the spectrum alone. These are the singlets **J** and **M**, and the doublets **F** and **G**. The singlets **J** and **M** are due to the methyl groups attached to the head of the molecule. The hydrogens from both these methyl groups are not within three bonds of any other hydrogens, and so no coupling is observed in the spectrum. As discussed above, the two methyl groups are in different environments and so give rise to two well-separated signals.

The signals **F** and **G** are perhaps hardest to understand. These signals are due to the –CH_2– groups in the head part of the structure. However, the hydrogens that are equivalent are the two hydrogens that are 'up' (one on the left-hand side, and one on the right), and the two hydrogens that are 'down' (again, one on the left, and one on the right). These are shown below:

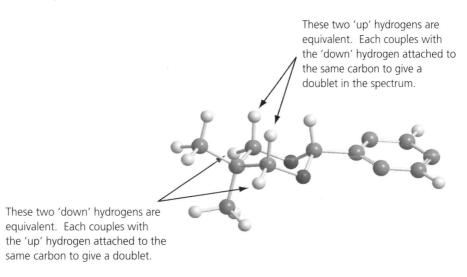

These two 'up' hydrogens are equivalent. Each couples with the 'down' hydrogen attached to the same carbon to give a doublet in the spectrum.

These two 'down' hydrogens are equivalent. Each couples with the 'up' hydrogen attached to the same carbon to give a doublet.

Within each equivalent –CH$_2$– group, the 'up' hydrogen couples with the down hydrogen, since they are different and separated by just two bonds. However, it is not possible to precisely assign which is which based on this spectrum alone.

The table should therefore be completed as follows:

^{1}H NMR signal	Hydrogen(s) on Carbon(s)
A	7
B	**19, 23**
C	9, 11
D	36
E	17
F	37, 39
G	37, 39
H	14, 3
I	15, 2
J	40 or 41
K	27, 28, 29, 33, 34, 35
L	16, 1
M	40 or 41

RSC | Advancing the Chemical Sciences

The synthesis of *Fexofenadine*

From Round 1 2008, Question 6.

Antihistamines are taken to reduce the effects of allergic reactions in the body. The drug *Fexofenadine* is used to treat sneezing, runny nose and itchy eyes experienced by hay fever sufferers, without causing drowsiness.

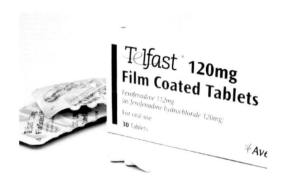

Fexofenadine, structure shown below, is sold as the hydrochloride salt with each tablet containing 112 mg of *Fexofenadine* as 120 mg of the salt with HCl.

(a) i)   On the structure of *Fexofenadine* below, circle the atom which will be protonated in the salt.

   ii)   Given that 112.00 mg of *Fexofenadine* is actually contained in 120.14 mg of the salt, calculate the relative molecular mass of *Fexofenadine*.

   iii) Given your answer to part (ii), suggest what common organic group R is in the structure of *Fexofenadine* below.

A synthesis of *Fexofenadine* is shown below.

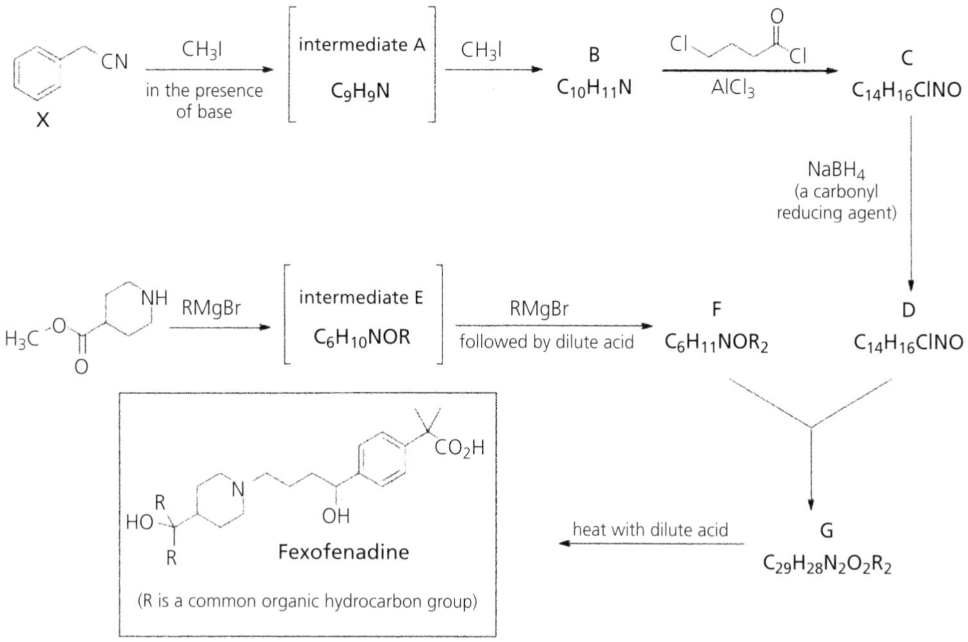

X → (CH₃I, in the presence of base) → [intermediate A C_9H_9N] → (CH₃I) → B $C_{10}H_{11}N$ → (AlCl₃) → C $C_{14}H_{16}ClNO$

C → (NaBH₄ (a carbonyl reducing agent)) → D $C_{14}H_{16}ClNO$

(piperidine methyl ester) → (RMgBr) → [intermediate E $C_6H_{10}NOR$] → (RMgBr followed by dilute acid) → F $C_6H_{11}NOR_2$

F + D → G $C_{29}H_{28}N_2O_2R_2$ → (heat with dilute acid) → Fexofenadine

Fexofenadine

(R is a common organic hydrocarbon group)

(b) Draw the structure of starting material **X** and indicate which hydrogen atoms are removed by the base in the first step of the synthesis.

(c) Draw the structures of compounds / intermediates **A** to **G**.

The synthesis of *Fexofenadine* – an analysis

(a) i) *Fexofenadine* is sold as a hydrochloride (HCl) salt. Looking at the structure of the drug molecule, it is the nitrogen atom of the amine group which will be protonated.

The structure of *Fexofenadine* is given to candidates in which two parts of the molecule have been abbreviated by an "R" symbol. If the molecular mass of the *Fexofenadine* molecule is known, then we can deduce which functional group "R" actually represents.

ii) We are told that 112.00 mg of the drug molecule are present in 120.14 mg of the salt, so the remainder of this mass (8.14 mg) must be due to HCl. We can therefore calculate the number of moles of HCl present.

$$\text{No. moles HCl} = \frac{\text{Mass of HCl}}{\text{Molecular mass (HCl)}} = \frac{8.14 \times 10^{-3}\,g}{36.5\,g\,mol^{-1}} = 2.23 \times 10^{-3}\,\text{moles}$$

Since *Fexofenadine* and HCl are present in a 1:1 molar ratio, this is also the number of moles of *Fexofenadine*. We are now in a position to calculate the relative molecular mass of *Fexofenadine*.

$$\text{Molecular mass (Fexofenadine)} = \frac{\text{Mass of Fexofenadine}}{\text{Number of moles}} = \frac{112.00 \times 10^{-3}\,g}{2.23 \times 10^{-3}\,\text{moles}} = 501.6\,g\,mol^{-1}$$

iii) The formula of *Fexofenadine* as drawn is $C_{20}H_{29}NO_4R_2$, with a relative molecular mass of 347.4 plus the mass of the two R groups. Since this needs to equal 501.6, the molecular mass of a single R group must therefore be 77.1. The identity of R is actually C_6H_5, namely a phenyl group.

Candidates are then asked to consider a synthetic scheme in which the structures of some of the intermediates are missing. The carbon skeleton of the target molecule *Fexofenadine* is shown, so even if some of the reactions are unfamiliar, this structure can be used as a helpful guide.

(b) Molecule **X** possess **H** atoms on the carbon atoms of the aromatic ring and also on the carbon adjacent to the nitrile group. If we locate the skeleton of **X** in the final target molecule, we see the two methyl groups which are introduced during this step have both been incorporated at this position and not at any of the ring positions. This also gives us an important clue as to which protons are removed by the base. In fact, this nitrile group stabilises the anion formed when these protons (marked) are removed, so these can be removed by base.

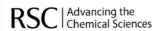

(c) As explained in part b) the base removes the protons from the carbon next to the nitrile, and the anion formed then reacts with the iodomethane. This can happen twice, since there are two protons that can be removed. From the molecular formulae, we see that each step to **A** and then **B** involves the overall addition of CH_2. Looking at the final target, the two methyl groups are both located at this position next to the aromatic ring.

X		**intermediate A**		**B**
C_8H_7N		C_9H_9N		$C_{10}H_{11}N$

From **B** to **C** involves the addition of C_4H_5OCl. Candidates may recognise the reagents as a Friedel-Crafts acylation of a benzene ring. Looking ahead to the structure of *Fexofenadine*, we can also see that the aromatic ring is disubstituted at opposite carbon atoms of the ring. Therefore there is no ambiguity over where the new group needs to be introduced. We can also tell that the oxygen atom ends up at the carbon atom adjacent to the ring.

intermediate B		**C**
$C_{10}H_{11}N$		$C_{14}H_{16}ClNO$

We are told that sodium borohydride ($NaBH_4$) reduces carbonyls, and the molecular formula is increased by two H atoms only. Therefore the ketone is reduced to a secondary alcohol in this step.

C		**D**
$C_{14}H_{16}ClNO$		$C_{14}H_{18}ClNO$

E is formed from the addition of R-MgBr to the ester shown. Such a reagent may have been encountered before as a *Grignard reagent*, which will attack carbonyls. The molecular formulae show a loss of CH_3O and an incorporation of R in forming **E**, and a further incorporation of R and an H atom in forming **F**. Looking ahead to the structure of *Fexofenadine* and locating the nitrogen-containing six-membered ring, we can see that the two R groups both end up in the position shown.

		Intermediate E		**F**
$C_7H_{13}NO_2$		$C_6H_{10}NOR$		$C_6H_{11}NOR_2$

The coupling together of molecules **D** and **F** produces **G**, which contains the same number of carbon atoms as *Fexofenadine*. Therefore the carbon skeleton should be identical to that of the final product. This reaction sees an amine attacking a chloroalkane in a *nucleophilic substitution* reaction.

To confirm the structure of **G**, we see that the final step involves the acid hydrolysis of the nitrile group to form the carboxylic acid of *Fexofenadine*.

The synthesis of Viagra™

From Round 1 2006, Question 6.

The reaction scheme shown overleaf is based on the first synthesis of sildenafil. This is the active ingredient in Viagra, the drug used for the treatment of "male erectile dysfunction".

Note that the by-products are not necessarily indicated in this scheme.

(a) Draw structures for the intermediates **B**, **D**, **E**, **F**, **G**, **I**, and **J**.

(b) i) Suggest suitable reagent(s) for the conversion of **B** to **C**.

    ii) Suggest a suitable reagent for the formation of **F** from **E**.

(c) In the formation of **I** from **H**, the first step in the mechanism is a deprotonation by the sodium hydroxide solution. On the answer sheet, indicate clearly which hydrogen is removed by the base.

(d) Deduce the structure for *N*-methylpiperazine, the reagent needed to convert **J** to sildenafil. Compound **A** is actually prepared by the reaction between hydrazine, N_2H_4, and reagent **K** according to the balanced equation below.

$$K \; + \; N_2H_4 \; \rightarrow \; A \; + \; 2\,H_2O$$

(e) i) Draw the structure for hydrazine.

    ii) Suggest a structure for **K**.

Reaction scheme based on the first synthesis of sildenafil

The synthesis of Viagra™ – an anlysis

We include an organic synthesis question in Round 1 each year. In this (and questions from other years) a step-wise sequence of chemical transformations are used to elaborate a simple starting molecule into the given target molecule.

Candidates are asked to deduce the structures of missing intermediates: the key to solving this puzzle is to pull together all of the information given in the question, and not to panic if any of the reagents are unfamiliar. We would encourage students to keep as much of the carbon framework in the structure as intact as possible and just look for new elements as they appear in the course of the synthesis.

(a) For familiar reagents it may be possible to deduce the product of a reaction with no additional information. However, for unfamiliar reagents the molecular formulae can be used to deduce the new atoms that have been incorporated into the structure. The position of these new atoms in the structure can often be found by looking at the structures that appear later on in the synthesis.

Intermediate B: The reagent may be unfamiliar, but from the molecular formulae of **A** and **B**, we can deduce that CH_2 has been added. From the structure of **C** we can see that one of the nitrogen atoms now bears a methyl group, so the reagent must methylate this nitrogen.

$C_9H_{14}N_2O_2$ (A) → $C_{10}H_{16}N_2O_2$ (B)

Intermediate D: From the formulae, NO_2 is added whilst H is lost in going from **C** to **D**. The combination of nitric acid and conc. sulfuric acid may have been met before as a nitrating agent. Looking forwards to structure **H**, we can see that a C-N bond is added to the pyrrole skeleton, so the nitration must occur at this position.

C $C_8H_{12}N_2O_2$

HNO$_3$
conc. H$_2$SO$_4$

(forms reactive species NO$_2^+$)

D $C_8H_{11}N_3O_4$

Intermediate E: From **D** to **E**, Cl is added whilst OH is lost. Sulfonyl chloride is routinely used to convert carboxylic acids to acid chlorides.

D $C_8H_{11}N_3O_4$

SOCl$_2$

E $C_8H_{10}N_3O_3Cl$

(+ SO$_2$ + HCl)

Intermediate F: From **E** to **F**, NH_2 is added whilst Cl is lost. Again, looking forwards to structure **H**, a formamide group $C=O(NH_2)$ is present where we currently have an acid chloride, which helps us to deduce the structure of **F**.

E $C_8H_{10}N_3O_3Cl$? **F** $C_8H_{12}N_4O_3$

Intermediate G: From **F** to **G**, two hydrogens are added whilst two oxygens are lost. **H** only differs by having an amide in place of our nitro group, and looking at the structure of the acid chloride reagent used in the next step of the synthesis we can deduce the structure of **G**. H_2 therefore reduces the nitro group to a primary amine.

F $C_8H_{12}N_4O_3$ H_2 / Pd **G** $C_8H_{14}N_4O$

We should emphasise that you do not have to work rigidly through the structures in alphabetical order. For example, if the structure of **F** is not yet known, it should still be possible to deduce the structure of **G** by working backwards from **H**. Removing the portion of the molecule which is clearly introduced by the acid chloride reagent would reveal the primary amine **G**.

Intermediate I: We are told that the dehydration (loss of H_2O) of **H** gives **I**. Glancing ahead to the structure of sildenafil, we see that a ring is formed from the two amides and that H_2O is lost from this part of the molecule. Even if we are unsure of the mechanism, there are enough other clues in the question to deduce the structure of **I**.

H $C_{17}H_{22}N_4O_3$ $-H_2O$ **I** $C_{17}H_{20}N_4O_2$

Intermediate J: From **I** to **J** incorporates SO_2Cl and loses H from the molecular formula. Looking at sildenafil tells us where the sulfur becomes attached to the molecule.

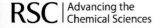

I $C_{17}H_{20}N_4O_2$ → J $C_{17}H_{19}SN_4O_4Cl$

(b) i) Having deduced the structure of **B**, candidates should recognise that the transformation of **B** to **C** involves the hydrolysis of an ester to a carboxylic acid. Ester hydrolysis can be performed under acidic or basic aqueous conditions.

ii) The transformation of **E**, an acid chloride, to **F**, an amide, requires the use of ammonia.

(c)   **proton removed here**

(d) Looking at the structure of sildenafil we see that there is a six membered ring containing two nitrogen atoms (one of which bears a methyl group). In the two steps going from **H** to **J**, none of the reagents used could possibly provide these atoms. Therefore the structure of *N*-methylpiperazine can be deduced.

J $C_{17}H_{19}SN_4O_4Cl$ — N-methylpiperazine → Viagra $C_{22}H_{30}SN_6O_4$

(e) i) The pyramidal structure of ammonia should be familiar, in which nitrogen has three bonds to hydrogen and a lone pair. In hydrazine, the two nitrogen atoms are pyramidal and are joined by an N-N bond.

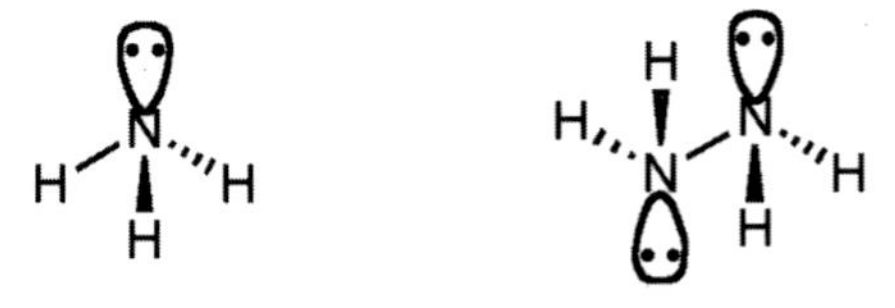

ammonia NH$_3$ hydrazine N$_2$H$_4$

ii)   The structure of **K** can be found working backwards from **A**, removing the two
nitrogen atoms that come from hydrazine. Since two molecules of water are
produced, the two oxygen atoms must be incorporated in the structure of **K**.

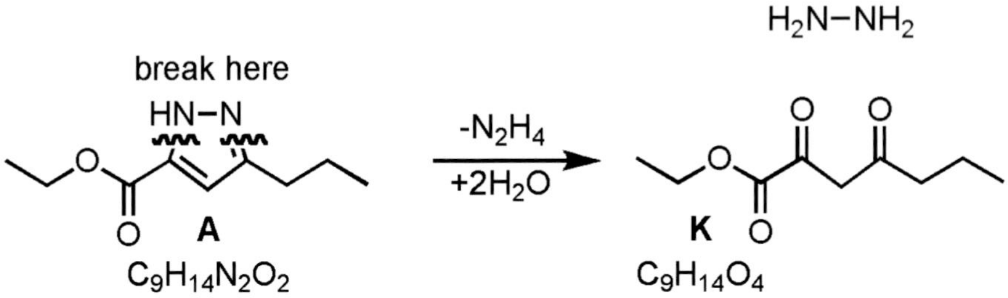

H$_2$N−NH$_2$

break here

HN−N

$\xrightarrow[+2H_2O]{-N_2H_4}$

A
C$_9$H$_{14}$N$_2$O$_2$

K
C$_9$H$_{14}$O$_4$

Toxins from cone snails

From Round 1 2008, Question 5.

Cone snails are predators that use venom to capture prey. The toxic species in the venom are polypeptides. Cone snail toxins are of pharmaceutical interest as starting points for the development of new anaesthetics. A number of research groups are working towards identifying the amino acid sequences of new cone snail toxins.

The cone snail

Polypeptides are polymers of amino acids; the structures and relative masses of some amino acids are shown below:

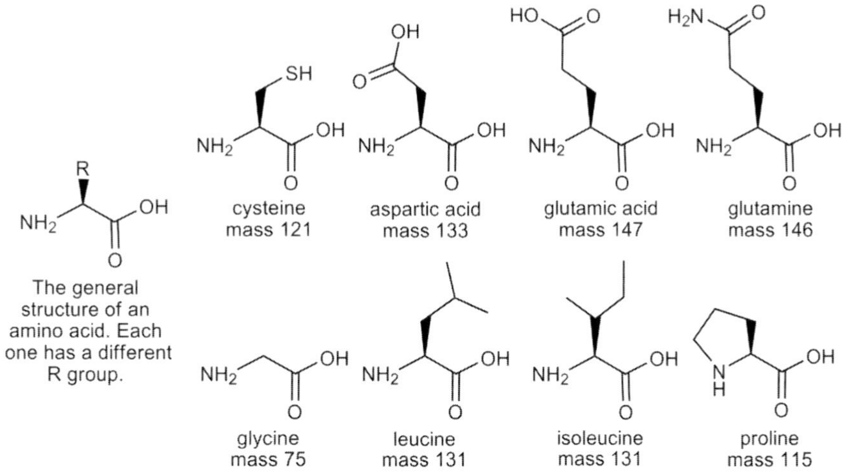

The general structure of an amino acid. Each one has a different R group.

cysteine
mass 121

aspartic acid
mass 133

glutamic acid
mass 147

glutamine
mass 146

glycine
mass 75

leucine
mass 131

isoleucine
mass 131

proline
mass 115

When amino acids form a polypeptide, an amide bond is made and water is lost:

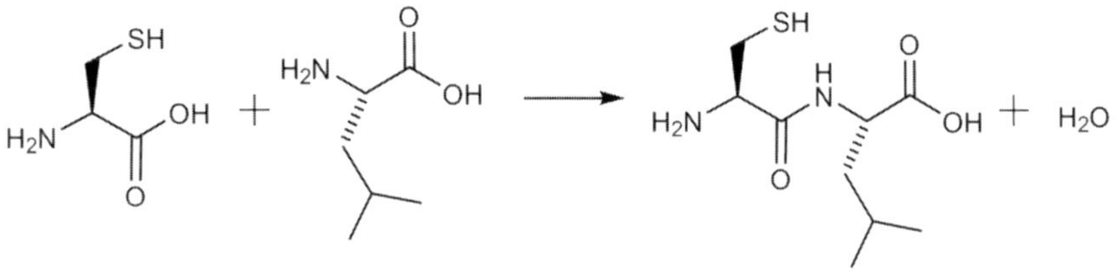

In biological systems the function of a polypeptide depends on the order of the amino acids in the sequence. By convention, a polypeptide is drawn starting with the amine group on the left, hence the sequence of the polypeptide shown above is cysteine-leucine NOT leucine-cysteine.

Polypeptides are often sequenced using mass spectrometry. In a mass spectrometer the polypeptide breaks into fragments with the amide bonds being the most likely to be broken. By comparing the masses of the different ions formed it is possible to work out the amino acid sequence. The major ions seen in the fragmentation of an isoleucine-leucine-glycine polypeptide are shown overleaf:

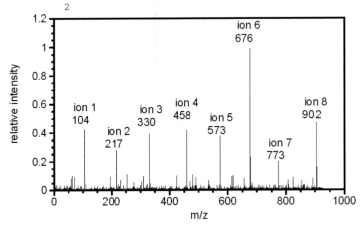

Ion 1   **Ion 2**

isoleucine-leucine-glycine polypeptide   Fragment ions seen in the mass spectrum

In all parts of this question you should use the mass of the most common isotopes of each of the elements: 12 for C, 14 for N, 16 for O and 1 for H.

(a) i) What is the mass of the isoleucine-leucine-glycine polypeptide?

ii) What is the mass of **Ion 1**?

iii) What is the mass of **Ion 2**?

A polypeptide, **X**, of mass 976 was isolated from a cone snail. It was found by chemical analysis to have the following amino acid composition:

2 x cysteine, 1 x aspartic acid, 1 x glutamic acid, 1 x glutamine, 1 x glycine, 1 x isoleucine, 1 x leucine and 1 x proline.

(b) How many unique polypeptide sequences can be formed using all these amino acids?

The fragmentation mass spectrum of **X** and the ^{1}H NMR spectrum of the third amino acid in the sequence are shown below. Under the conditions used for the NMR spectrum no peaks are seen for the NH_2 and COOH protons.

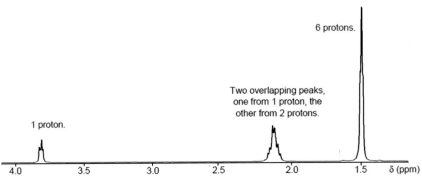

(c) The last two amino acids in the sequence are glutamic acid-glycine. What is the sequence of the first 7 amino acids? [The structures and masses of the amino acids are shown at the top of page 41.]

Toxins from cone snails – an analysis

This biological chemistry question involves the interpretation of graphical data. The polypeptide molecules may be unfamiliar to many candidates but there are hints in the early parts of the question that should help with the solution of later sections.

(a) It is possible to do all parts of section (a) by finding the number of each different type of atom in each compound and then multiplying by the appropriate atomic mass. In many cases this method gives errors, in particular it is easy to make a mistake counting the number of protons in each compound. An alternative approach is to use the masses and structures of the amino acids given in the question. This is also helpful for solving later sections of the problem.

  i)   When a polypeptide is formed from two amino acids a water molecule is lost, so: M_r (two amino acid polypeptide) = M_r (amino acid 1) + M_r (amino acid 2) – M_r (H_2O). Every time an additional amino acid is added to the polypeptide another water molecule is lost, therefore:

M_r(isoleucine-leucine-glycine polypeptide) = M_r(isoleucine) + M_r(leucine) + M_r(glycine) – (2 x $M_r(H_2O)$)

  ii) The mass of Ion 1 can be found by comparison with the structure given for isoleucine:

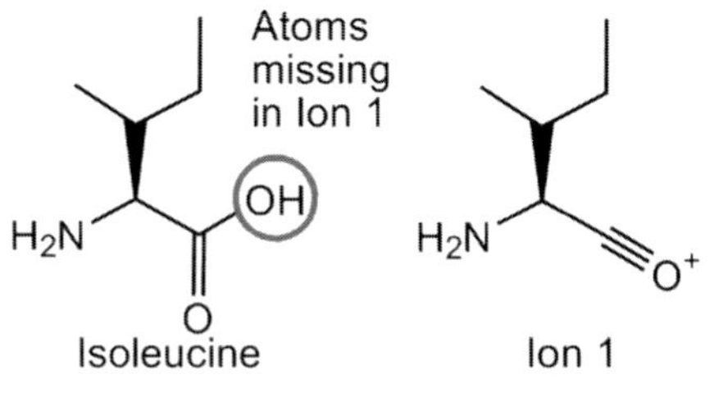

M_r (Ion 1) = M_r (isoleucine) – M_r (OH)

Isoleucine   Ion 1

  iii) In a similar way to part (ii) the mass of Ion 2 can be found by comparison with the structure of an isoleucine-leucine polypeptide:

Isoleucine-leucine polypeptide   Ion 2

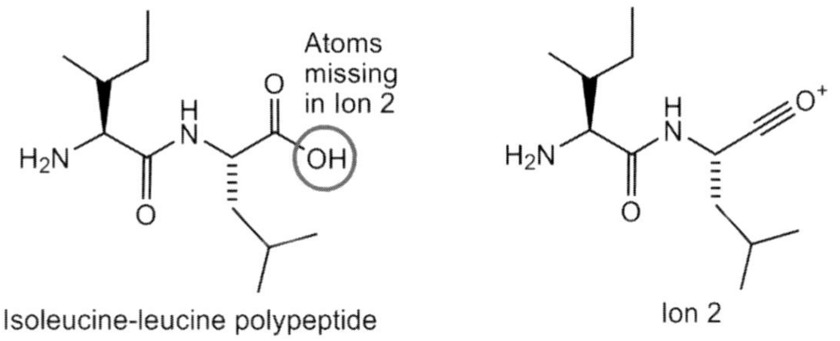

(b) The number of different permutations is 9! (because there are 9 amino acids – 9! is 9 x 8 x 7 x 6 x 5 x 4 x 3 x 2 x 1) divided by 2! (the number of different ways of arranging the two cysteine amino acids). This gives 181440 different sequences!

(c) The fragmentation mass spectrum of the 9 amino acid polypeptide and the identity of the last two amino acids in the sequence are given in this section. One way to approach this problem is to look at which bonds are broken in the fragmentation of the isoleucine-leucine-glycine polypeptide from section (a):

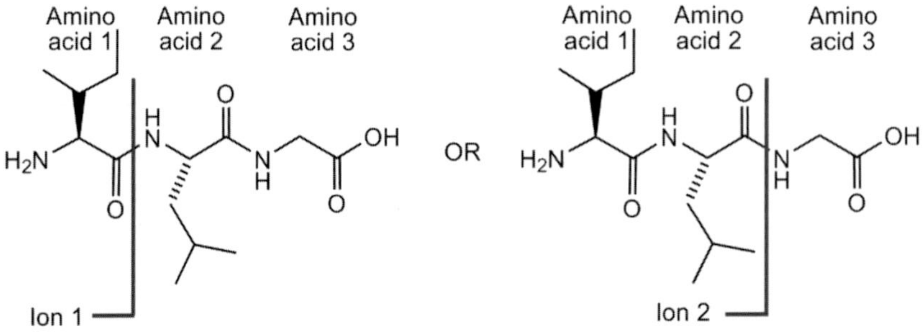

From section (a) it is known that M_r (Ion 1) = M_r (amino acid 1) – M_r (OH), therefore the first amino acid in the sequence of the 9 amino acid polypeptide must be cysteine (M_r = 121). To find the identity of the remaining amino acids the difference in mass between consecutive ions can be used:

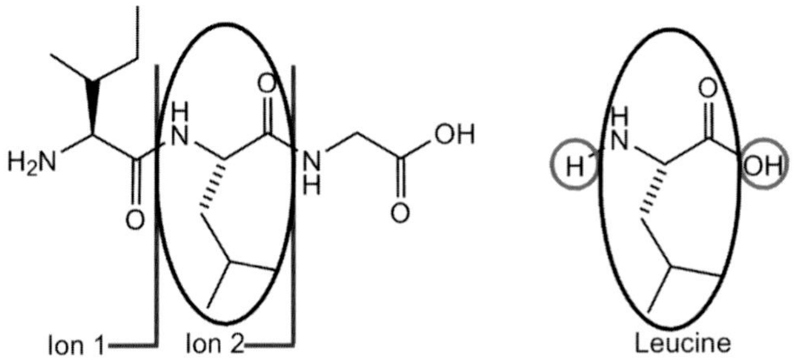

Comparing the regions circled in black shows that:
M_r (Ion 2) – M_r (Ion 1) = M_r (amino acid 2) – M_r (H$_2$O)

Finding the differences in mass between consecutive ions allows the amino acid sequence to be determined as: 1, cysteine, 2, isoleucine or leucine, 3, isoleucine or leucine, 4, glutamine, 5, aspartic acid, 6, cysteine and 7, proline (amino acids 8 and 9 are given).

Finally, the NMR spectrum can be used to differentiate between isoleucine and leucine at position 3. The number of peaks in the ^{1}H NMR spectrum will correspond to the number of different ^{1}H chemical environments. It is important to note that the question states that neither the alcohol nor amine protons will be visible under the conditions used to take this spectrum. Protons in different chemical environments are indicated in the structures below:

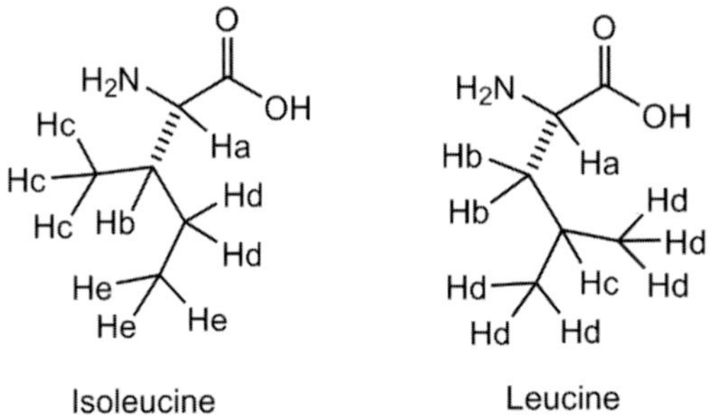

The ^{1}H NMR spectrum given has four peaks (two of which overlap), and is therefore the spectrum of leucine. Amino acid 2 is therefore isoleucine whilst 3 is leucine.

Aluminium chemistry and rat poison

From Round 1 2008, Question 4.

Aluminium metal reacts with various non-metals to form simple, binary compounds. The reaction with phosphorus forms aluminium phosphide, AlP. This compound has been used as a rodenticide.

The type of bonding in aluminium compounds depends on which element it is bonded to. For example, aluminium oxide is predominantly ionic, whereas aluminium chloride (empirical formula $AlCl_3$) shows characteristics of covalent bonding.

(a) How many outer electrons are around each Al atom in a covalently bonded $AlCl_3$ molecule?

In the vapour phase at 150-200 °C, aluminium chloride exists as a molecule, **A**, which has an M_r of 266.66.

(b) i)   What is the molecular formula of **A**?
    ii) Suggest a structure for **A**.
    iii) How many outer electrons are around each Al atom in your structure of **A**?

Aluminium phosphide is hydrolysed by water to generate the highly toxic gas phosphine, PH_3. Phosphine is similar in structure to ammonia, and like NH_3, PH_3 can act as a ligand using its lone pair.

(c) Write an equation for the hydrolysis of AlP.

There has been interest in various compounds containing Al-P covalent bonds as precursors for AlP. When equal moles of iBu_2AlH and Ph_3SiPH_2 are dissolved in solvent at 25 °C, hydrogen gas is evolved and a white crystalline solid **B** is produced ($iBu = (CH_3)_2CHCH_2-$; $Ph = C_6H_5-$).

(d) How many outer electrons are around Al in the covalently bonded iBu_2AlH?

The mass spectrum of **B** gives a peak with highest m/z value at 864.

(e) i)   Using your answer to (b) as a guide, suggest a structure for compound **B**.
    ii) Compound **B** shows isomerism. Draw structures to indicate the three-dimensional shape of two geometric isomers of **B**.

When warmed, **B** is converted to **C** with the evolution of methylpropane. The ^{31}P–NMR spectrum of **C** showed it to have a single environment for phosphorus, and ^{13}C–NMR showed equal numbers of iBu- and Ph₃Si- groups. Further analysis showed the compound to have four Al and four P atoms in the molecule.

(f) Suggest a structure for compound **C**.

When **C** is heated to temperatures above 150 °C, it starts to decompose, yielding Ph₃SiH and a gas **D**. By 500 °C, all that remains is aluminium phosphide.

(g) Identify the gas **D**.

Aluminium chemistry and rat poison – an analysis

This question was essentially about the ability of aluminium to accept a dative bond from an electron pair donor to complete its octet in the outer shell. The first half of the question (parts (a)-(d)) was designed to get you thinking about this area of chemistry using aluminium chloride as the model. Most, if not all, A-level specifications require candidates to be able to rationalise the chemistry of $AlCl_3$ in terms of its structure and bonding, and therefore it was assumed that the structure of Al_2Cl_6 should be known to most Olympiad candidates. Even if it were not, it was hoped that you would follow the train of thought below:

(a) There are only 6 electrons in the outer shell of Al in a molecule of $AlCl_3$ and this doesn't therefore conform to the octet rule.

(b) i) The M_r of the gaseous form of aluminium chloride suggests a molecular formula of Al_2Cl_6.

     ii) The structure of Al_2Cl_6 probably, therefore, allows the Al atoms to complete their octet, the only source of "extra" electrons being the lone pairs on the Cl atoms. Clearly, there are not enough electrons in the system to allow Al–Al bonds, which was a common answer. So the answer to (b) part (ii) was:

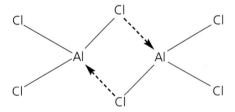

     iii) This question simply confirms that the structure now has 8 electrons in the outer shell of each Al atom.

The question now points out that phosphine, PH_3, is like ammonia and can act as a ligand through the lone pair of electrons on the P atom. This is an important piece of information – P can form co-ordinate (dative covalent) bonds.

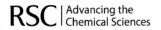

(c) The equation for the hydrolysis of AlP to phosphine should have been pretty straightforward – we have asked lots of this type of question in Round 1 in the past. The side product could be either $Al(OH)_3$ or Al_2O_3.

Now we get to the main thrust of the question. The two compounds iBu_2AlH and Ph_3SiPH_2 are introduced.

(d) First you are asked how many electrons are in the outer shell of Al in iBu_2AlH – the answer being 6, ie it is like $AlCl_3$, in that it is electron deficient (does not have a full outer shell).

Then we turn to the reaction product **B** and its structure. The mass spectrometry data for **B** should have helped you. The M_r of iBu_2AlH is 142 and that of Ph_3SiPH_2 is 292. Having been told that **B** is formed from equal moles of these two compounds, with the evolution of hydrogen gas, and it has an M_r of 864, this should have led you to spot that $2 \times (142+292) = 868$, and so **B** must be formed from two each of iBu_2AlH and Ph_3SiPH_2 with the evolution of 2 moles of H_2.

(e) i) Now, given the question so far (dative covalent bonds to Al allowing it to complete its octet, P being a lone pair donor) you should have brought all this information together to come up with the following answer:

The Al-P bonds are formed by the loss of an H atom from each, which results in the two molecules of H_2 being evolved as **B** is formed. The dative bonds P→Al are there to complete the octet of the Al atoms.

ii) This compound can form geometric isomers (a form of stereoisomerism: same molecular formula, same structural formula, different arrangement of bonds in space) because the presence of the 4-membered ring prevents the rotation of any of the Al-P bonds. This means that the two $SiPh_3$ groups can be on the same side of the ring (cis) or opposite sides (trans). This could have been indicated in your diagrams with conventional wedge and dotted bonds.

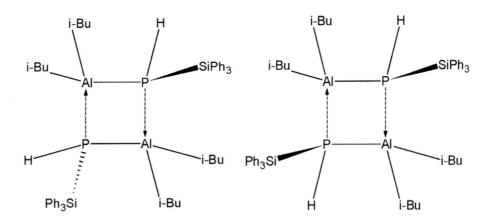

(f) This is perhaps the most challenging part, but, given what has gone before, it should have been possible. The spectroscopic data for **C** tells you that the structure is very symmetrical. There is only a single environment for phosphorus. But in the structures that you have so far drawn for **B**, there are twice as many *i*Bu groups as there are Ph$_3$Si groups. You are told that in **C**, there are equal numbers of these. This, taken together with the information that methylpropane is evolved as **C** is formed (*ie i*Bu + H) and there are now 4 Al and 4 P atoms per molecule of **C**, should have led you to the rather beautiful structure that is given below:

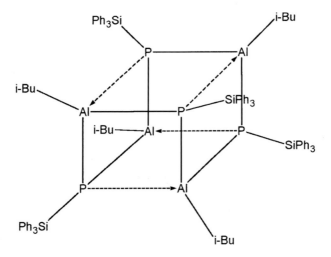

As you can see, **C** is formed by bringing together two molecules of **B**. Think of them as the top and bottom faces of the cube. The two faces are joined by Al-P covalent bonds as a result of the elimination of methylpropane.

(g) When **C** is heated to destruction, AlP remains, as Ph$_3$SiH is given off, together with the gas **D**. Therefore **D** must be C$_3$H$_6$ or methylprop**ene**, that is *i*Bu minus an H.

This was a difficult question, but it was designed to give the thinking chemist all of the information that they needed to crack the problem. It did require careful reading, but the chemistry is all approachable by a good A-level candidate. What we always try and do in Olympiad Round 1 questions is to take A-level knowledge and stretch it into some interesting, and in this case beautiful, areas.

The thermal decomposition of copper(II) sulfate

From Round 1 2006, Question 4.

Thermogravimetry is an analytical technique which involves heating a substance and measuring the change in mass.

The graph below shows the change in mass as copper(II) sulfate pentahydrate, $CuSO_4.5H_2O$, is heated.

Decomposition takes place where the gradient is steepest leaving various decomposition products indicated by **A** to **F** in the diagram.

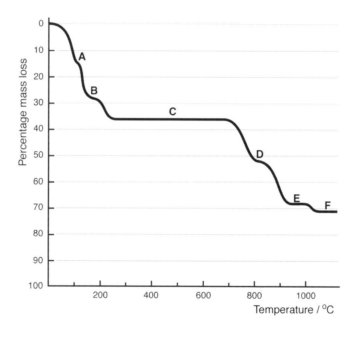

(a) Using the data from the graph, suggest formulae for compounds **A**, **B**, and **C**.

(b) On heating **E**, a redox reaction occurs to form **F**. Identify **E** and **F** and write an equation for this reaction.

(c) Compound **D** is formed when exactly half of **C** has decomposed to form **E**. What is the empirical formula of **D**?

The thermal decomposition of copper(II) sulfate – an analysis

This is a challenging question involving interpretation of graphical data in addition to chemical knowledge gained from A-level studies. Using graphical data and drawing appropriate conclusions is a vital part of any scientific career, both for analysing experimental results and for reviewing published data.

The question gives a graph showing the change in mass as copper(II) sulfate pentahydrate, $CuSO_4.5H_2O$, is heated. The graph can be used to find the chemical formulae for decomposition products formed at different temperatures – these products are labelled A to F and are represented by horizontal regions on the graph. The first step in answering this question is to use the graph to estimate the percentage of the mass of $CuSO_4.5H_2O$ that is lost in forming each of the compounds A to F. It is also helpful to calculate the percentage of the mass of $CuSO_4.5H_2O$ lost between consecutive species and the mass of $CuSO_4.5H_2O$ (249.7).

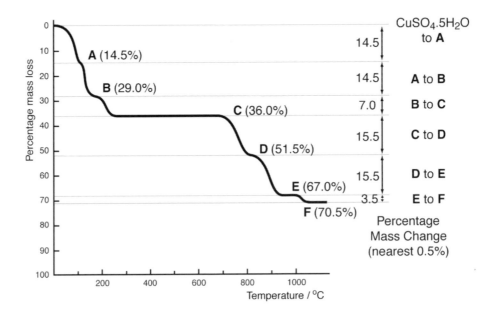

(a) Formation of compounds **A**, **B** and **C** gives a reduction in mass relative to $CuSO_4.5H_2O$ of 14.5%, 29% and 36% respectively. At this point you should use your chemical knowledge and think about what species are likely to be lost from $CuSO_4.5H_2O$ on heating. Consider the type of bonds present and how much energy will be required to break them. With these factors in mind it is most likely that the water molecules will be lost first. The loss of one mole of water from one mole of $CuSO_4.5H_2O$ corresponds to a 7.2% reduction in mass. Starting with one mole of $CuSO_4.5H_2O$ this suggests that the formation of **A** would correspond to the loss of 2 moles of water, **B** to the loss of 4 moles of water and **C** to the loss of 5 moles of water. Therefore **A** is $CuSO_4.3H_2O$, **B** is $CuSO_4.1H_2O$ and **C** is $CuSO_4$.

The structure of $CuSO_4.5H_2O$ has been determined using x-ray and neutron diffraction. Each copper ion has an octahedral coordination with four water molecules in the equatorial positions and an oxygen atom from a sulfate ion in each of the axial positions. The fifth water molecule does not coordinate to copper but instead interacts with the sulfate ions. The structure of $CuSO_4.5H_2O$ is illustrated online at **http://www-teach.ch.cam.ac.uk/links/crystals/web/CuSO4-2.html**.

(b) A redox reaction occurs when compound **F** is formed from compound **E**. There are two common oxidation states for copper, copper(II) and copper(I), so the redox reaction will involve the formation of a copper(I) species. Compound **E** can be identified by looking at the percentage of the mass of $CuSO_4.5H_2O$ lost between compounds **C** and **E** in addition to the formula for compound **C**. At this point it is useful to calculate the percentage mass that one mole of Cu, S and O atoms contribute to the mass of one mole of $CuSO_4.5H_2O$ (25.5%, 12.8% and 6.4% respectively). From the graph it can be seen that approximately 31% of the mass of $CuSO_4.5H_2O$ is lost between compound **C** and compound **E**. This is consistent with the loss of one mole of S atoms and three of O atoms (32.1% mass loss), or of one mole of Cu atoms with one mole of O atoms (31.9%). Only the first of these options is sensible – it results in the loss of species that we would expect to be a gas at 800 °C and gives CuO as the formula for compound **E**.

In forming compound **F** from compound **E** (CuO), approximately 3.5% of the mass of $CuSO_4.5H_2O$ is lost. Looking at the mass change we would expect for losing a mole of O atoms (6.4%), this suggests that half of the O atoms are lost when compound **F** is formed. This gives $CuO_{1/2}$, properly written as Cu_2O, as the formula for **F**. This is copper(I) oxide – so the formation of **F** from **E** is a redox reaction as required. The equation for this reaction is:

$$2\ CuO(s) \rightarrow Cu_2O(s)\ +\ \tfrac{1}{2}\ O_2(g)$$

(c) Finally, compound **D**, is formed half way between the decomposition of compound **C** ($CuSO_4$), to compound **E** (CuO). In forming one mole of compound **E** from a mole of compound **C** 1 mole of S and 3 moles of O atoms are lost. Therefore halfway through this process 0.5 moles of S and 1.5 moles of O atoms are lost. Subtracting this from the formula for **C** gives $CuS_{1/2}O_{5/2}$, which should be written Cu_2SO_5.

A supernova

From Round 1 2004, Question 6.

The electronic ground state (*ie* the lowest electronic state) of a hydrogen atom may be written $1s^1$ indicating that the single electron resides in the 1s orbital. If sufficient energy is given to the atom, the electron may be promoted from the 1s orbital to a higher energy orbital, such as the 2p orbital or the 3p orbital.

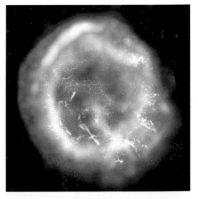

Supernova remnant E0102-72 as photographed by the UV / x-ray telescope *Chandra* . (NASA)

The energy of an electron in a hydrogen atom (or any ionized atom with nuclear charge Z and with just one electron remaining) is given by the following equation:

$$E_n = -R_H \frac{Z^2}{n^2}$$

The energy of a free, ionized electron is zero; electrons in the atom have lower energy, hence the minus sign.

In the equation, Z is the number of protons in the nucleus ($Z = 1$ for hydrogen);
n is the principal quantum number ($n = \mathbf{1}$ for the **1**s orbital, **2** for the **2**s and **2**p orbitals, **3** for the **3**s, **3**p and **3**d orbitals, *etc.*);
R_H is the *Rydberg constant* equal to the ionization energy of a hydrogen atom ($R_H = 2.179 \times 10^{-18}$ J).

(a) Calculate the energy of an electron in a 2p orbital in an excited hydrogen atom.

(b) Calculate the energy needed to promote the electron in a hydrogen atom from the 1s tal to the 2p orbital.

(c) Calculate the ionization energy of a helium ion, He^+.

When an electron returns from a higher energy orbital to a lower one, energy is given out as light (the cause of the familiar flame colours). The frequency of the light, f, (in Hz) is related to the energy of the transition, ΔE, by the equation:

$$\Delta E = hf \quad \text{(where } h \text{ is Planck's constant} = 6.626 \times 10^{-34} \text{ J s)}$$

(d) Calculate the frequency of light for the electronic transition in a hydrogen atom from a 2p orbital to the 1s orbital (the so-called Hydrogen Lyman-α line).

Supernova remnant E0102-72, located some 200,000 light years away, has been found to

contain more than a billion times the amount of oxygen contained in the Earth's oceans and atmosphere. At the incredibly high temperatures in the supernova (many millions of Kelvin), the oxygen atoms are multiply ionized to single electron species, O^{7+}. The oxygen was detected by the specific frequency of its Lyman-α line (the transition $n = 2$ to $n = 1$).

(e) Calculate the frequency of the O^{7+} Lyman-α line.

(f) Another element present in large quantities has its Lyman-α line at 2.471×10^{17} Hz. What element is this?

A supernova – an analysis

The equation for the energy, E_n, of an electron in a hydrogen atom, or in any ionized atom with nuclear charge Z and with just one electron remaining is given in the question as:

$$E_n = -R_H \frac{Z^2}{n^2}$$

where Z is the number of protons in the nucleus, n is the principal quantum number, and R_H is the Rydberg constant (equal to the ionization energy of a hydrogen atom; $R_H = 2.179 \times 10^{-18}$ J).

Looking at this equation, we note that for a given ion (*ie* with Z fixed), the energy gets less and less negative, tending towards zero, as the principal quantum number, n, increases. This corresponds to the electron being in higher and higher energy shells and located, on average, further away from the nucleus. At zero energy, the electron is no longer associated with the nucleus.

For the electron in the atom or ion, the energy is always negative. This is relative to the zero energy that the free electron possesses when it is completely separated from the nucleus. To remove the electron from the atom or ion requires us to put in energy to raise its energy up to zero, hence in the atom or ion, the energy of the electron must be negative.

Note that the equation implies that, for example, the energy of an electron in a 2s orbital is exactly the same as that of an electron in a 2p orbital. Likewise, the energy of an electron in a 3s, 3p, or 3d orbital would be the same. This is because the energy for a one-electron system depends only on the principal quantum number of the shell and not on which sub-shell it is in. This result may seem odd since we are used to statements such as "the 2s sub-shell is lower in energy than the 2p sub-shell" but this statement is only actually true for systems with more than one electron (which includes all atoms other than the hydrogen atom).

(a) For an electron in a 2p orbital ($n = 2$) in a hydrogen atom (where $Z = 1$), the energy, E_2, is given by:

$$E_2 = -2.179 \times 10^{-18} \times \frac{1^2}{2^2} = -0.5448 \times 10^{-18} \text{ J}.$$

As expected, this is higher in energy (ie less negative) than for an electron in the 1s orbital.

(b) Care needs to be taken over the sign for this answer, although it should be intuitive.

Energy needed to promote the electron from the 1s to the 2s

= (energy for the electron in the 2s) − (energy for the electron in the 1s)

= (-0.5448×10^{-18} J) − (-2.179×10^{-18} J) = + 1.634×10^{-18} J.

As we can see, the answer is positive, as expected, meaning energy needs to be put in to promote the electron to a higher energy level.

(c) To ionize a He^+ ion ($Z = 2$), the electron must be raised from the 1s shell ($n = 1$) to zero energy. The energy of the electron in the 1s shell is:

$$E = -2.179 \times 10^{-18} \times \frac{2^2}{1^2} = -8.716 \times 10^{-18} \text{ J}.$$

Therefore the ionization energy is + 8.716×10^{-18} J.

(d) From (b), the energy given out, ΔE, when an electron returns from the 2p orbital in hydrogen to the 1s orbital is +1.634×10^{-18} J.

The frequency, f, that corresponds to light of this energy is given by:

$$f = \frac{\Delta E}{h} = \frac{1.643 \times 10^{-18}}{6.626 \times 10^{-34}} = 2.480 \times 10^{15} \text{ Hz}$$

This could be converted to a wavelength, λ, using the relationship $\lambda = c / f$, where c is the speed of light (3.00×10^8 m s^{-1}). The wavelength comes out to be 1.210×10^{-7} m or 121.1 nm. This is in the UV region of the electromagnetic spectrum.

The data for the following part of the question originates from the Chandra X-ray observatory. A search for

supernova E0102-72

should reveal many beautiful colour photographs of this supernova on the NASA website.

(e) For the single-electron O^{7+} ion, $Z = 8$. The energy change during the transition from $n = 2$ to $n = 1$ is given by:

(energy for the electron in the 1s) − (energy for the electron in the 2s)

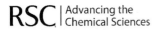

$$= \left(-2.179\ 10^{-18}\ x\ \frac{8^2}{1^2} \right) - \left(-2.179\ x\ 10^{-18}\ \frac{8^2}{2^2} \right) = -1.046\ x\ 10^{-16}\ J.$$

Note that the energy change is negative meaning energy is given out during this transition.

The frequency of light corresponding to this energy is

$$f = \frac{\Delta E}{h} = \frac{1.046\ x\ 10^{-16}}{6.626\ x\ 10^{-34}} = 1.579\ x\ 10^{17}\ Hz.$$

This is in the X-ray region of the electromagnetic spectrum.

(f) A frequency of 2.471×10^{17} Hz corresponds to an energy of

$$(2.471 \times 10^{17}) \times (6.626 \times 10^{-34})\ J = 1.637 \times 10^{-16}\ J.$$

We can set up the following equation for the Lyman-α line (the transition from $n = 2$ to $n = 1$) for the unknown ion with atomic number Z:

$$= \left(-2.179\ 10^{-18}\ x\ \frac{Z^2}{1^2} \right) - \left(-2.179\ x\ 10^{-18}\ \frac{Z^2}{2^2} \right) = -1.637\ x\ 10^{-16}.$$

Rearranging this gives:

$$= \left(\frac{Z^2}{1} \right) - \left(\frac{Z^2}{4} \right) = \frac{3Z^2}{4} = \frac{-1.637\ x\ 10^{-16}}{-2.179\ x\ 10^{-18}}$$

$$\text{So}\ Z = \sqrt{\frac{4\ x\ (1.637\ x\ 10^{-16})}{3\ x\ (2.179\ x\ 10^{-18})}} = 10$$

The element with atomic number $Z = 10$ is neon.

References

References for questions

Q3   "The Crystal and Molecular Structure of Mercury Fulminate".
W.Beck et al., Zeitschrift für anorganische und allgemeine Chemie, 2007, 633,
1417–1422.

Q6   "Synthesis of Anthropomorphic Molecules: The NanoPutians"
Chanteau and Tour, Journal of Organic Chemistry, 2003, 68(23), 8750–8766.

Q7   "A new synthesis of carboxyterfenadine (*Fexofenadine*) and its bioisosteric tetrazole
analogs"
B. Di Giacomo et al., Il Farmaco, 1999, 154, 600-610.

Q8   "Sildenafil (VIAGRA^{TM}), a potent and selective inhibitor of type 5 cGMP
phosphodiesterase with utility for the treatment of male erectile dysfunction"
Nicholas K Terrett, Andrew S Bell, David Brown and Peter Ellis, Bioorganic &
Medicinal Chemistry Letters, 1996, 6, 1819-24.

Q9   "A vasopressin/oxytocin – related conopeptide with g-carboxyglutamate at position 8"
Carolina Möller and Frank Marí, Biochemical Journal, 2007, 404, 413-419.

Q10   "New Aluminium Phosphide Precursor"
Alan H. Cowley et al., Angewandte Chemie International Edition, 1990, 29, 1409–
1410.

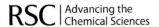